Collecting, Exploring, and Interpreting Microbiological Data Associated with Reclaimed Water Systems

About the WateReuse Foundation

The mission of the WateReuse Foundation is to conduct and promote applied research on the reclamation, recycling, reuse, and desalination of water. The Foundation's research advances the science of water reuse and supports communities across the United States and abroad in their efforts to create new sources of high quality water through reclamation, recycling, reuse, and desalination while protecting public health and the environment.

The Foundation sponsors research on all aspects of water reuse, including emerging chemical contaminants, microbiological agents, treatment technologies, salinity management and desalination, public perception and acceptance, economics, and marketing. The Foundation's research informs the public of the safety of reclaimed water and provides water professionals with the tools and knowledge to meet their commitment of increasing reliability and quality.

The Foundation's funding partners include the Bureau of Reclamation, the California State Water Resources Control Board, the Southwest Florida Water Management District, the California Energy Commission, and the California Department of Water Resources. Funding is also provided by the Foundation's Subscribers, water and wastewater agencies, and other interested organizations.

Collecting, Exploring, and Interpreting Microbiological Data Associated with Reclaimed Water Systems

A Guidance Manual

Audrey D. Levine
U.S. Environmental Protection Agency

Valerie J. Harwood
University of South Florida

Gordon A. Fox
University of South Florida

Cosponsors

Bureau of Reclamation
California State Water Resources Control Board

Published by the WateReuse Foundation
Alexandria, VA

Disclaimer

This report was sponsored by the WateReuse Foundation and cosponsored by the Bureau of Reclamation and the California State Water Resources Control Board. The Foundation, its Board Members, and the project cosponsors assume no responsibility for the content reported in this publication or for the opinions or statements of facts expressed in the report. The mention of trade names of commercial products does not represent or imply the approval or endorsement of the WateReuse Foundation, its Board Members, or the cosponsor. This report is published solely for informational purposes.

For more information, contact:
WateReuse Foundation
1199 North Fairfax Street, Suite 410
Alexandria, VA 22314
703-548-0880
703-548-5085 (fax)
www.WateReuse.org/Foundation

WateReuse Foundation Project Number: WRF-04-012
WateReuse Foundation Product Number: 04-012-01

ISBN: 978-1-934183-17-5
Library of Congress Control Number: 2009925042

Printed in the United States of America

♻ Printed on Recycled Paper

CONTENTS

TABLES

FIGURES

EXAMPLES

FOREWORD

The WateReuse Foundation, a nonprofit corporation, sponsors research that advances the science of water reclamation, recycling, reuse, and desalination. The Foundation funds projects that meet the water reuse and desalination research needs of water and wastewater agencies and the public. The goal of the Foundation's research is to ensure that water reuse and desalination projects provide high-quality water, protect public health, and improve the environment.

A Research Plan guides the Foundation's research program. Under the plan, a research agenda of high-priority topics is maintained. The agenda is developed in cooperation with the water reuse and desalination communities including water professionals, academics, and Foundation Subscribers. The Foundation's research focuses on a broad range of water reuse research topics including:

- Definition and addressing of emerging contaminants;
- Public perceptions of the benefits and risks of water reuse;
- Management practices related to indirect potable reuse;
- Groundwater recharge and aquifer storage and recovery;
- Evaluation and methods for managing salinity and desalination; and
- Economics and marketing of water reuse.

The Research Plan outlines the role of the Foundation's Research Advisory Committee (RAC), Project Advisory Committees (PACs), and Foundation staff. The RAC sets priorities, recommends projects for funding, and provides advice and recommendations on the Foundation's research agenda and other related efforts. PACs are convened for each project and provide technical review and oversight. The Foundation's RAC and PACs consist of experts in their fields and provide the Foundation with an independent review, which ensures the credibility of the Foundation's research results. The Foundation's Project Managers facilitate the efforts of the RAC and PACs and provide overall management of projects.

The Foundation's primary funding partners include the Bureau of Reclamation, California State Water Resources Control Board, the Southwest Florida Water Management District, the California Energy Commission, Foundation Subscribers, water and wastewater agencies, and other interested organizations. The Foundation leverages its financial and intellectual capital through these partnerships and funding relationships.

This guidance manual is the result of a Foundation-sponsored research study. The focus of the manual is on how to use statistical tools to answer questions about the prevalence, removal, or survival of microorganisms in the context of wastewater reclamation and reuse. The information provided in this manual is intended to help with routine monitoring programs and design of studies for detailed microbiological investigations.

David L. Moore
President
WateReuse Foundation

G. Wade Miller
Executive Director
WateReuse Foundation

ACKNOWLEDGEMENTS

This project was funded by the WateReuse Foundation in cooperation with the Bureau of Reclamation and the California State Water Resources Control Board. This project was conducted at the University of South Florida as a collaborative effort through the Department of Biology and the Department of Civil and Environmental Engineering. The authors conducted an interdisciplinary course encompassing biology and engineering on water reuse in conjunction with the project, and several students provided input for the statistical topics, including Asja Korajkic, Phoebe Koch, Wendy Mussoline, Robert M. Ulrich, Cecilia Claudio, and Jennifer Perone. The authors appreciate the assistance of Jennifer Lynette in producing the final document.

The authors extend their thanks to the utilities that provided data on their monitoring and operations. Input from the utilities helped to frame the statistical questions explored in this project.

Participating Utilities and Organizations
City of Clearwater, Clearwater, FL
Jacksonville Electric Authority (JEA), Jacksonville, FL
Sanitation Districts of Los Angeles County, Los Angeles, CA
King County, Seattle, WA

The authors thank the WateReuse Foundation, the project managers, and the project advisory committee (PAC) members for helping to steer this guidance document. The PAC provided excellent feedback throughout the project. The project manager was Burnett King. PAC members included

Katie Benko, M.S., *Bureau of Reclamation*
George D. DiGiovanni, Ph.D., *Texas A&M University*
Michael Messner, *U.S. Environmental Protection Agency*
Rich Mills, *California State Water Resources Control Board*
Vanessa Speight, Ph.D., P.E., *Malcom Pirnie, Inc.*

Project Team
Audrey D. Levine, *University of South Florida, during the project, currently with the U.S. Environmental Protection Agency*
Valerie J. Harwood, *University of South Florida*
Gordon A. Fox, *University of South Florida*

EXECUTIVE SUMMARY

This guidance manual provides users with a context for collecting, exploring, and interpreting microbiological data associated with reclaimed water. Basic concepts are presented to facilitate collection of meaningful data with an emphasis on the design of sampling programs, data interpretation, and statistical analysis. The information provided in this manual is intended to help with routine monitoring programs and design of studies for detailed microbiological investigations. The examples and illustrations encompass a variety of microbial investigations relevant to reclaimed water facilities.

Designing statistically sound sampling programs for microbiological testing of reclaimed water requires consideration of several interrelated factors but must start with determination of the specific goal(s) or monitoring question(s) to be answered, including the microbiological analytes to be measured. This manual provides readers with an overview of sources of error associated with sampling and analysis. It also addresses the challenge of balancing resource limitations and the availability of analytical methods with the complex array of factors that influence the microbial characteristics of reclaimed water. Key issues associated with sampling that are relevant for reclaimed water studies, including the type of sample (grab, composite, continuous, or online), sampling location (before or after treatment or distribution), sample volume, timing, and hydraulic considerations, are presented. The importance of collecting the appropriate number of samples is illustrated through several examples focused on reclaimed water applications and statistically based questions.

A straightforward approach for statistical manipulation of microbial data is provided using a series of examples that range in complexity. The manual first introduces descriptive statistics and then describes basic calculations such as calculating log reduction and dealing with data sets that include nondetected values. Information and examples are provided to help the user understand statistical sources of error and the way in which careful study design facilitates interpretation of data. The manual also provides insight into the significance of "significance." The importance of power analysis in designing and interpreting studies is presented using a series of examples. An overview of hypothesis testing is provided with examples of paired and unpaired data, univariate and multivariate analysis of variance, correlation analysis, and binary logistic regression.

The manual is not a textbook but is an effort to demystify some of the challenges of statistically based analyses. The users are encouraged to seek additional resource materials when necessary, but we hope that this manual provides the basis for exploring statistical considerations for the average user, who may have previously shied away from a subject that can be intimidating. The main "take-home" messages are

- Frame your questions carefully, with consideration of available supporting data and resources.

- Plan your experimental design with system characteristics and specific goals in mind; for example, are analyses to be paired or unpaired, can many samples be analyzed, or are there cost considerations that will necessitate compromises?

- Obtain preliminary data; carry out descriptive statistical analysis and make graphs to gain an understanding of the variability and other characteristics of the data.

- Determine the expected necessary sample size given the variability and the magnitude of the difference you wish to detect.

- Have fun with statistical calculations and interpretation of results! Don't be afraid to try using different statistical approaches. Once you start exploring data and seeing relationships within and among data sets, it can become an absorbing and very helpful part of your skill set.

CHAPTER 1

INTRODUCTION

Ensuring the microbiological safety of reclaimed water systems and protecting end users from exposure to pathogens require constant vigilance of reclaimed water quality through process and operational controls coupled with sound monitoring programs. Ideally, microbial sampling programs should yield information about the prevalence of pathogens and potential health risks associated with water reuse applications. In reality, microbial sampling efforts are constrained by resource limitations and the availability of accurate, cost-effective, and efficient methods for pathogen monitoring. Therefore, it is important to understand how the design of sampling programs affects the quality of data and our ability to interpret the results effectively. This guidance document provides practical information on the application of statistical tools for planning microbial testing programs, interpreting data, and communicating the results and conclusions. Recommendations for using statistical tools for retrospective analysis of historical data are also provided.

The focus of this manual is on how to use statistical tools to answer questions about the prevalence, removal, or survival of microorganisms in the context of wastewater reclamation and reuse. It is not intended to be an exhaustive discussion of statistical practices; rather, it should be a practical guide to framing and answering questions about microbial aspects of water reuse practices through correct application of basic statistics. Practical concepts of data analysis are presented that can be used to help defend decisions about process performance and operational efficiency. This can be achieved by applying statistical procedures to determine if variations in data are due to "random chance" or correlated to other factors such as wastewater sources and characteristics, precipitation events, flowrates and loading rates, or the effectiveness of treatment systems. The importance of discriminating between statistical significance and biological significance is also explained. The manual provides guidelines to help determine the types of statistical tests to use for answering specific questions, including testing assumptions about the data, testing hypotheses, and interpreting data. Throughout the document, examples of the statistical concepts and applications are presented using a case study approach.

CHAPTER 2

SAMPLING PROGRAM DESIGN

Designing statistically sound sampling programs for microbiological testing of reclaimed water requires consideration of several interrelated factors but must start with determination of the specific goal(s) or monitoring question(s) to be answered, including the microbiological analytes to be measured. The success of a sampling program depends on taking the appropriate types of samples at the right location at the right time and frequency. While the majority of microbial sampling of reclaimed water is conducted for compliance monitoring, designing microbial sampling programs to incorporate supporting data such as co-analyses of physical-chemical water quality parameters and process information can allow for a more comprehensive and informative interpretation of test results. An important aspect of sampling for microbial analytes is to select an appropriate methodology to address the goals of the testing program. Analytical approaches vary in cost, turnaround time, and the ability to yield quantitative information on microbial concentrations, speciation, viability, and infectivity. When information is needed on the concentrations of viable organisms, it is important to understand relationships between the volume of sample that is collected and the detection limit. In general, the concentrations of viable microorganisms decrease with each successive stage of reclaimed water treatment. Therefore, the volume of sample needed depends on the extent of treatment, the specific analyte(s) to be tested, and the goals of the sampling program (e.g., compliance, troubleshooting, investigative studies, process optimization). In addition, the presence of particulate matter may necessitate supplemental processing of samples to quantify pathogens that may be associated with suspended particles or biofilms.

2.1   DEFINITION OF SAMPLING PROGRAM GOALS

Microbiological characterization of reclaimed water can be conducted to answer specific questions such as

- Is the disinfected effluent in compliance with regulatory requirements?
- To what extent is treatment (individual treatment units or an entire treatment train) effective for reduction of pathogens?
- Are there seasonal or short-term variations in the degree to which microorganisms persist or survive through treatment or distribution? Can this variability be explained by process, loading, and/or water quality information?
- Do changes in disinfection strategies (pre-and post-treatment, contact time, dosage, etc.) and/or the disinfectant affect survival of pathogens?
- Do rainfall events and hydraulic conditions affect pathogen persistence through wastewater treatment processes?
- Do the relative concentrations of indicators and pathogens change with storage and distribution of reclaimed water?

It is important to understand the details of the treatment operations and hydraulics prior to defining the appropriate question(s) to be answered. Details about sampling programs that should be planned to meet the program goals include

- the specific analytes to be considered (e.g., coliform bacteria, adenoviruses, *Acanthamoeba, Cryptosporidium, Giardia*, other indicators, etc.) and the potential for using surrogates and/or indicators in conjunction with (or in lieu of) pathogen testing,
- the degree of quantification needed (presence or absence, microbial concentrations, speciation, viability, infectivity, etc.),
- the level of accuracy, interference, and detection limits associated with quantifying each analyte,
- the specific sampling location(s),
- the timing and frequency of sample collection,
- the minimum number of samples needed for statistical purposes, and
- the type of supporting documentation needed (Table 2-1).

Examples of specific microbiological sampling program goals are listed in Table 2-1. The relationship among sampling goals, decision variables, and other influencing factors is shown in Figure 2-1. Analysis of the data collected allows interpretation and communication of the results.

Table 2-1. The goals of a sampling program affect both the sampling approach and the supporting documentation needed

Goal of Sampling Program	Considerations for Sampling Approach	Supporting Documentation Needed	Examples of Sampling Program Goals and Questions
Compliance monitoring	Sample location, parameters	Flowrate, disinfection parameters	Daily indicator organism monitoring (e.g., coliforms) Routine pathogen monitoring
Assessment of process performance	Sample location, parameters, replicates	Flowrate, process information, plant hydraulics, and water quality	Are filters removing protozoan pathogens?
Optimization of process	Sample location, parameters, replicates, frequency	Flowrate, process information, plant hydraulics, and water quality	Does flowrate affect filter performance for removal of protozoan pathogens?
Troubleshooting	Sample location, parameters, replicates, frequency, timing	Flowrate, process information, maintenance history, plant hydraulics, and water quality	What factors contribute to sporadic detection of fecal coliforms in reclaimed water?

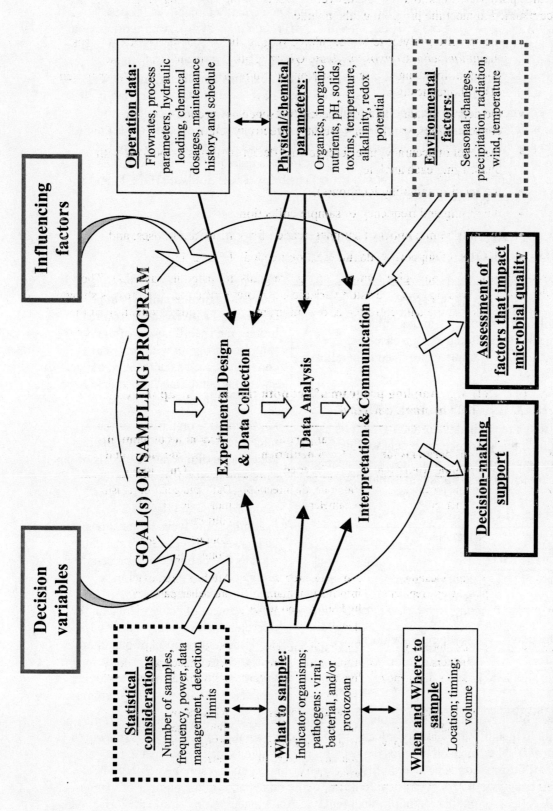

Figure 2-1. The interrelationships between the goals of the sampling program and the factors that influence experimental design and data collection.

5

The U.S. Environmental Protection Agency (USEPA) has developed guidance on the planning process and on criteria for data acceptability in environmental studies in a document entitled "Guidance on Systematic Planning using the Data Quality Objectives Process" (http://www.epa.gov/QUALITY/qs-docs/g4-final.pdf) (*Guidance*, 2006). An important goal of study planning is to ensure that available resources are used to collect the data that will be most useful to answering the questions at hand and that the study design is such that meaningful information will be obtained. A good example of this consideration is power analysis, which can calculate the sample size needed to have a certain level of certainty that the answer is correct, given the known variability in the data. A key component of study planning is determining the measures that will be taken to ensure that data are of sufficient quality to provide accurate and meaningful information. The USEPA has also designed software to aid in study design, particularly with respect to data quality objectives. The software is called Data Quality Objectives Error Feasibility Trials Software (DEFT) and is available at http://www.epa.gov/quality/qa_docs.html.

2.2 WHAT SHOULD BE CONSIDERED IN SELECTING MICROBIOLOGICAL ANALYTES?

CASE STUDY

Examples of the concepts in this guidance document are illustrated using a case-study approach. Each example is provided in parallel with the text.

BOX 1: SITE OF THE STUDY

Data from a reclaimed water facility will be used throughout this guidance document to illustrate key concepts. An overview of the plant capacity and treatment stages is given in Table CS-1.

Table CS-1. Case Study Overview

Parameter	Details
Capacity (MGD)	150
Biological treatment	Biological nutrient removal
Filtration	Dual media
Disinfection	Sodium hypochlorite
Indicator monitoring	Total coliform

Selection of the appropriate suite of microbiological parameters to be analyzed in a given study requires linking the specific goals of the sampling program with practical constraints such as personnel, budget, and turnaround time (how quickly are the results needed?).

A systematic comparison of the microbiological analytical tools that are capable of meeting the goals of the sampling program should be conducted in the early phases of sampling program design. For example, some types of questions can be answered using categorical data such as detect/nondetect (presence or absence) and/or above/below a specific compliance level. Development of categorical data for microbial analytes is frequently less costly than generation of quantitative data and therefore may provide an opportunity to introduce additional replicates into the sampling program.

To answer other types of questions, quantitative data may be needed. In some cases it may be important to quantify the viability/infectivity of the microorganisms (e.g., *Cryptosporidium* and *Giardia*), whereas in other cases the use of total cell counts may be adequate. For example, if information is needed on filtration performance, changes in the number of cells (total cell counts) will reflect physical removal, while differences in the number of viable cells will reflect a combination of physical removal AND inactivation. Thus, categorical data such as the presence or absence of indicators or pathogens may be less useful than quantitative data. Similarly, to assess health risks, knowledge about cell viability is much more important than information on total cell numbers (viable plus nonviable). Alternatively, rapid semiquantitative or qualitative measurements can serve as "triggers" for more intensive microbial analysis.

In some cases surrogate parameters, such as particle counts, suspended solids, or turbidity, can be used to provide indirect measures of physical removal of microorganisms; however, while these parameters can supplement direct microbial measurements, they cannot provide information on viability. Historically, microbiological testing programs for reclaimed water have relied upon routine testing of indicator organisms because of the relative availability of analytical methods and regulatory requirements. Data on indicator organisms may be supplemented by intermittent testing of pathogens, depending on the overall goals of the investigation. Examples of indicator organisms that have been used for testing of reclaimed water are listed in Table 2-2. Note that in general, particulate matter shields microorganisms from the effects of physical and chemical disinfectants. Typically, microorganisms that are classified as indicator organisms (e.g., coliforms, *Escherichia coli*, enterococci, etc.) are predominately nonpathogenic and are associated with sources of fecal material. Coliform bacteria (total, fecal, or *E. coli*) are the most widely used indicator organisms for reclaimed water applications. The terminology used to refer to the various groups of coliform bacteria can be confusing and is explained in Table 2-3.

In contrast to indicator organisms, pathogenic organisms in reclaimed water may be associated with sources of fecal material but also may be present in reclaimed water from other sources or opportunistic growth within storage and distribution systems. Another important distinction between indicators and pathogens is that, if pathogenic organisms are present in reclaimed water, they have the potential to cause disease, depending on their effective concentration, and the likelihood of exposure occurring through inhalation, ingestion, and/or skin contact. For pathogens that have been studied in reclaimed water (enteric viruses, *Giardia*, *Cryptosporidium*), reported concentrations tend to be orders of magnitude lower (<10/100 L) than the concentrations of indicator organisms (>1/100 mL) and their occurrence is more intermittent, depending on the health status of the community served by a treatment facility. It should also be noted that pathogen testing is about 20 to 100 times more expensive than testing of indicator organisms; therefore, prudent planning of pathogen testing is critical to ensure that appropriate resources are available to collect meaningful data.

The concept of testing indicator organisms in lieu of testing pathogens is based on the paradigm that the sources, fate, transport, growth characteristics, and survival patterns associated with indicator organisms have some parallel relationship to the properties of pathogens. In reality, while pathogens and indicators may co-occur in many environments (e.g., sewage), they differ in physical size, surface characteristics, chemical resistance, and other factors that affect their concentrations and persistence through the physical, biological, and chemical treatment technologies associated with the production of reclaimed water. Consequently, the ratio of pathogens to indicator organisms can vary with each stage of treatment because of relative differences in microbial susceptibility and vulnerability to removal mechanisms. Therefore, it is important that microbiological parameters are selected to address the goals of the sampling program where practicable. Reliance on monitoring a single indicator, while convenient, is not adequate for evaluation of pathogen survival. For example, in a study of six reclaimed water facilities, five out of 25 reclaimed water samples contained infectious *Cryptosporidium* oocysts, yet fecal coliforms were not detected in any of the samples that contained infectious oocysts (<0.2 colony-forming units [CFU]/100 mL) (Harwood et al., 2005).

Table 2-2. Comparison of indicator organisms that can be used to assess microbiological quality of reclaimed water

Indicator organisms	Description	Relevance to reclaimed water monitoring	Relationship to pathogen persistence through reclaimed water production
Heterotrophic plate count (HPCs)	Total number of viable heterotrophic bacteria. Includes coliforms, enterococci, *Clostridium* spp., and other pathogenic and nonpathogenic heterotrophic bacteria.	Used as conservative measure of the survival of bacteria through reclaimed water production. Not widely used for testing reclaimed water.	No direct relationship. Although pathogens are also heterotrophic, HPC data are a gross indicator of the bacteria that are culturable under these conditions (e.g., aerobic, high nutrients), which is unrelated to the presence or absence of pathogens.
Coliform bacteria	Rod-shaped, gram-negative, non-spore-forming bacteria that have the ability to ferment lactose with gas production within 48 h at 35 °C (see Table 2-3 for more details on the coliform group).	Widely used to monitor disinfected effluents.	No direct relationship. Less resistant to disinfectants than are some protozoan pathogens (e.g., *Giardia* and *Cryptosporidium*).
Enterococcus spp.	*Enterococcus* spp. are a gram-positive, catalase-negative subgroup of fecal streptococci that tolerate a wide temperature range (10–45 °C), high ionic strength (6.5% NaCl), and elevated pH (9.5).	Used in monitoring of beaches, not widely used for monitoring of reclaimed water.	No direct relationship. Susceptibility to disinfectants is slightly greater than that of coliforms because of cell wall structure and tolerance of a range of ionic strength values, temperatures, and pH values.
Clostridium perfringens	*C. perfringens* bacteria are gram-positive, spore-forming, obligate anaerobic rod-shaped inhabitants of the intestines of warm-blooded animals. Endospores can resist desiccation, radiation, starvation, and disinfection.	Not widely used for testing of reclaimed water.	No direct relationship; however, the endospores are more resistant to disinfectants than other bacterial indicators.
Coliphages	Viruses that infect *E. coli*. Somatic coliphages attach to receptors on the cell wall of *E. coli*, and F-specific (F+) phages attach to the sex pilus of *E. coli*. Examples: f2, T₂, MS2, and ΦX174. Genomes are either RNA or DNA.	Used as a surrogate for virus monitoring for some applications (e.g., UV disinfection).	Weak relationship. Coliphages are similar in size to some viral pathogens, but their relative susceptibility to disinfection varies with the type of coliphage, type of virus, disinfectant characteristics, and water quality (pH, temperature, turbidity, etc.).

8

Table 2-3. Definitions of terms used to describe indicator bacteria belonging to the coliform group

Terminology	Description	Relevance to reclaimed water monitoring	Other comments
Total coliforms	Rod-shaped, gram-negative, non-spore-forming bacteria that can ferment lactose with gas production in 48 h at 35 °C. Present in soil, vegetation, sediment, natural waters, and the feces of humans and animals. Examples: *E. coli*, *Klebsiella*, *Enterobacter*, and *Citrobacter*.	Monitoring of disinfected effluents required by some regulatory agencies (e.g., California, Washington, etc.).	Used to assess drinking water quality; nonspecific indicators of contamination from many sources including soil, stormwater, and sewage.
Fecal coliforms or thermotolerant coliforms	Thermotolerant subgroup of the total coliforms that can grow at 44.5 °C. Examples: *E. coli* and certain members of the genera *Klebsiella* and *Citrobacter*.	Monitoring of disinfected effluents required by some regulatory agencies (e.g., Florida, Arizona, etc.).	Shed in the feces of most warm-blooded animals and some cold-blooded ones. Many states use them as indicators of surface (recreational) water quality.
E. coli	*E. coli* is a bacterial species that is a member of the fecal coliform group and the total coliform group. A distinguishing feature of most strains is the presence of a particular enzyme that gives a "MUG+" phenotype.[a] Pathogenic strains can cause acute gastroenteritis, hemolytic-uremic syndrome, and other health effects (*E. coli* O157:H7).	Used in lieu of fecal coliform testing by some utilities.	Once believed to specifically indicate human sewage; now known to belong to the normal flora of the gastrointestinal tract of humans and many animals. A majority of strains form a symbiotic relationship with the host. Some strains of *E. coli* are pathogenic; however, *E. coli* levels are not correlated with the presence or absence/inactivation of pathogens.

[a]The enzyme β-glucuronidase mediates the cleavage of substrates such as 4-methylumbelliferyl-β-D-glucuronide, a synthetic molecule that fluoresces when it is acted upon by the enzyme.

9

CASE STUDY

BOX 2: MICROBIOLOGICAL ANALYTE CONSIDERATIONS

The county has decided to upgrade a water reclamation facility by replacing its existing filtration system with the goal of improved removal or reduction of pathogens. Resources have been allocated to test the microbial quality of the reclaimed water. The testing program is to be designed to evaluate if control of pathogens can be estimated by testing microbial indicators to assess treatment performance. The county is specifically interested in assessing the performance of alternative filtration units and also in determining the microbial quality of the reclaimed water.

Questions that need to be answered are

- **What** combination of microbial analytes should be tested?

- **Where** should samples be collected?

- **When** should samples be collected?

- **How many** samples are needed to allow for valid interpretation of the data?

- What **supplemental/supporting information** is needed?

- What **statistical tests** should be conducted?

- **How** should the data be reported and represented?

2.2.1 Analytical methods

Methods for microbiological characterization of reclaimed water vary in sensitivity, specificity, detection limits, turnaround time, and expense. The majority of the microbial tests conducted on reclaimed water consist of culture-based methods, which rely on selective-differential media to promote the growth of specific groups of microorganisms (e.g., coliforms, enterococci, viruses) over a specific time period (e.g., 24 to 48 h for bacteria and weeks for viruses). The target microorganisms are frequently quantified by counting the number of CFU or plaque-forming units (PFU) in the case of bacteriophages and some human viruses. One approach for quantifying microorganisms is coupling a multiple tube dilution series with cell culture on selective-differential agar to yield a statistical estimate known as the most probable number. For some applications, cell growth is quantified through turbidity changes or pH changes due to gas production. Details on analytical methods can be found in a variety of sources including the USEPA (http://www.epa.gov/nerlcwww/index .html), a standard industry reference (*Standard Methods*, 2005), and other microbiological manuals (see reference list in Appendix 3).

Other approaches for quantifying microorganisms include microscopic techniques (light, fluorescence, phase-contrast, etc.) and molecular techniques that detect or quantify nucleic acid (DNA or RNA). A comparison of the applications and limitations of different approaches for quantifying microorganisms is given in Table 2-4. Combining culture-dependent techniques with molecular methods and/or microscopic methods can yield a more complete characterization of reclaimed water microbiology, depending on the goals and constraints of the sampling program.

Table 2-4. Comparison of applications and limitations of methods for quantifying microorganisms in reclaimed water

Method	Description	Applications	Limitations
Culture dependent	Identification of viable and culturable[a] microorganisms	Heterotrophic plate count, Indicator bacteria (total coliforms, fecal coliforms, *E. coli*, enterococci, *C. perfringens*), coliphages, viruses, many bacterial pathogens	Detection limits, turnaround time, do not detect viable but nonculturable (VBNC) organisms without special methodologies
Microscopy	Use of differential/fluorescence stains to identify and quantify microorganisms using light or fluorescence microscopes	Total number of bacterial cells (fluorescence stain, e.g. DAPI stain), total viable cells (e.g., membrane potential stains), gram-negative or gram-positive bacteria, protozoan pathogens (immunofluorescence stains)	Concentration of cells, interference from other particulate matter, no information about viability/infectivity, lack of specificity
Molecular techniques	Identification of microorganisms based on genetic characteristics	Can identify presence or absence of specific microorganisms and quantify in some cases (e.g., quantitative PCR, also known as Q-PCR)	No information on viability or infectivity unless testing is coupled with cell culture; relatively expensive and requires high level of expertise

[a]When stressed by unfavorable environmental conditions, many bacterial species can enter a state that is analogous to dormancy in which they cannot be cultured on the normal media used to isolate them but remain viable and may be resuscitated when conditions are more favorable. This state is often called viable but nonculturable (VBNC).

2.2.2 Sampling goal considerations

An important aspect of design of sampling programs is to select assays that are appropriate for meeting the sampling goals. For example, if the effectiveness of disinfection is to be tested, the conclusions can be influenced by differences in the degree to which individual organisms are susceptible to the disinfection conditions (disinfectant, concentration, contact time, presence of particulate matter, etc.) and the ability to differentiate viable organisms from nonviable ones. By application of supplemental information on disinfectant susceptibility to interpretation of data gathered on gram-negative (e.g., coliform) bacteria, gram-positive (enterococcal) bacteria, viruses, and protozoa (*Amoeba, Cryptosporidium, Giardia)*, a robust study design can be developed. It is important that the response of one type of microorganism (i.e., coliform bacteria) does not necessarily imply that the response of other microbial groups (i.e., protozoan pathogens or viruses) can be predicted. In other situations, the need to evaluate viability is less relevant. For example, if reduction of microorganisms through filtration or membranes is to be evaluated, then the degree of association of microorganisms with particulate matter and the total number of microorganisms may be of more importance than the viability of individual microbial species.

Some approaches that can be applied to select microbial parameters in the context of sampling program goals are given in Table 2-5.

Table 2-5. Relationship between sampling program goals and selection of microbiological parameters to be measured

Goal of sampling program	Criteria for selecting microbiological parameters	Examples
Compliance monitoring	Permit requirements	Coliform bacteria: total coliforms, fecal coliforms or *E. coli*
		Coliphages: MS2
		Enteroviruses and protozoan pathogens
Assessment of process performance, process optimization, trouble shooting, process design, pilot studies	Expected concentration of microorganisms, turnaround time, resistance or susceptibility to removal mechanisms, resource limitations (personnel, funds)	Identify microorganisms with various levels of resistance to the treatment. Reliance on a single parameter may yield misleading results.
Optimization of process	Sample location, parameters, replicates, frequency, volume	Identify microorganisms that are detectable and have various levels of resistance to the treatment.
Troubleshooting	Sample location, parameters, replicates, frequency, timing	Identify microorganisms that are detectable and can be measured with a fairly short turnaround time. Verify with traditional measurements.

2.2.3   Sampling parameter selection

Selection of the types of parameters that are tested is a critical step in the overall design of a microbiological investigation. There are several important trade-offs to be considered:

- potential for conducting multiple microbial tests,
- costs,
- turnaround time,
- accuracy, and
- the statistical validity of testing.

CASE STUDY

BOX 3: SELECTION OF MICROBIOLOGICAL ANALYTES

The county has elected to test a suite of microbial analytes for this project. The parameters to be tested and the rationale for their selection are given in Table CS-3.

To keep analytical costs within the allocated budget, indicators will be used to assess general characteristics of the reclaimed water. Pathogen testing will be conducted for the filter comparison and on the disinfected effluent.

For the sample problems in this document, we will use data from total coliform testing and *Cryptosporidium* analyses.

Analyte	Rationale
Indicators	
Total coliforms	State regulatory requirement (daily monitoring)
Coliphages	Potential surrogate for viruses (more numerous and less expensive to analyze)
Pathogens	
Human enteroviruses, *Cryptosporidium*, *Giardia*	Need a direct measure of the presence or absence of pathogens in reclaimed water

While testing of indicator organisms is less expensive than testing of pathogens, there is no direct relationship between the presence or absence of indicators and the presence or absence of pathogens. In addition, microorganisms vary in their ability to survive through reclaimed water treatment because of differences in size, surface characteristics, association with particulate material, and the degree of susceptibility or resistance to chemical or photochemical disinfection.

The starting point for selection of parameters is to generate a list of potential analytes that are appropriate for addressing the objectives of the study and conduct a review of the attributes of the available analytical methods (detection limits, turnaround time, costs, etc.). If possible, it is worthwhile to conduct preliminary sampling to estimate analyte concentrations and identify potential sampling challenges and matrix interference.

2.2.4 Differences between grab and composite samples

Another important consideration is the type of sample to be collected for microbiological analysis. Typically, reclaimed water quality is characterized using either grab samples or composite samples. Grab samples reflect the characteristics of the water at the time and location of sampling. Thus, the characteristics of a sample collected under peak flow conditions may differ from samples collected under other flow conditions. Precipitation events can also impact water characteristics in cases where infiltration or inflow can result in short-term changes in flowrates or in facilities served by combined sewers. The intensity and duration of storm, coupled with antecedent events (dry conditions and time interval since previous storm) and land-use patterns (pervious vs. impervious surfaces, agriculture, and roadways) and local stormwater management practices, influence the net impacts of storm events on water samples. Thus, caution should be exercised in extrapolating information from different types of sampling situations. In contrast, composite samples consist of a series of grab samples collected over a longer time (typically 24 h) with each incremental sample proportional to either the flowrate or the time period sampled. In general, while composite samples dampen the effects of short-term flow or loading variations, they are not recommended for microbial sampling because of the potential for growth, die-off, or predation within the sampling device and other sources of contamination. These issues also render statistical analyses problematic.

Data from continuous online monitoring can be used to optimize the selection of the sampling time and provide supportive information. Examples of online monitoring tools that may be available at reclaimed water facilities include flowrate, pH, chlorine residual, turbidity, suspended solids, dissolved oxygen, oxidation-reduction potential, conductivity, and the use of online UV absorbance or transmission patterns as indirect measures of organic content. The advantage of online monitoring is that it provides "real-time" information on the characteristics of the reclaimed water at a specific point in treatment. Changes in process performance that are reflected by increases in turbidity, suspended solids, or organics may indirectly indicate changes in microbial concentrations in the reclaimed water. Development of methods for online quantification of microorganisms is an active area of research and, ultimately, may improve our ability to characterize microbiological water quality in real time.

In conjunction with the collection of samples for microbiological characterization, it is important to characterize the treatment operations and water quality associated with each sampling event. The type of information needed depends on the sample location and the goals of the sampling program. Examples of supplemental water quality and process data that are relevant to sampling locations in wastewater reclamation facilities are given in Table 2-6.

Table 2-6. Supplemental water quality and process data relevant to microbiological sampling programs corresponding to different sampling locations within a reclaimed water production facility

Sample location	Supplemental water quality and process data
Untreated wastewater	Flowrate, temperature, biological oxygen demand (BOD), total suspended solids (TSS), nitrogen, industrial waste contributions, potential toxicity
Biological treatment units	Flowrate, MLSS, MCRT (SRT), BOD, TSS, turbidity, dissolved oxygen, nitrogen, phosphorus, metals, toxins
Filtration	Flowrate, HLR, time since last backwash, BOD, TSS, turbidity, TOC, nitrogen, phosphorus
Disinfection	Flowrate, disinfectant demand (oxidant demand), disinfectant dose and residual, contact time, TOC, nitrogen, TSS, turbidity, particle count
Distribution systems	System hydraulics (pressure, flowrates, and turnover time), disinfectant residuals, turbidity or suspended solids, electron acceptors (dissolved oxygen, nitrite, nitrate, sulfate, iron, and manganese), nutrients, potential cross-connections, system integrity, operational practices (e.g., flushing)

2.3 WHAT SHOULD BE CONSIDERED IN IDENTIFYING SAMPLING LOCATIONS?

The specific location where reclaimed water samples are collected for microbiological testing should be directly related to the sampling program goals. However, in some cases, the characteristics of a sample or sampling location can introduce potential interference and bias in microbiological tests. For example, particulate material can mask or shield some microorganisms, leading to suboptimal or inaccurate results. Sloughing of biofilms from surfaces (tank walls, pump tubing, and sample ports) can introduce an additional sporadic source of microorganisms into the sample that can impact quantitative results, depending on what is being measured, the sample volume, and the number of replicates. The presence or absence of oxygen (or other electron acceptors) can impact the growth and survival patterns of some microorganisms, thereby modifying the distribution of viable microorganisms and inadvertently biasing sample interpretation. In addition, the presence of residual disinfectants, such as chlorine, can cause supplemental inactivation of microorganisms between the time of sample collection and sample analysis, resulting in lower microbial concentrations than those found in the original sample or changes in population dynamics by favoring the growth of opportunistic organisms or disinfectant-resistant organisms. Conversely, quenching of residual disinfectants (using thiosulfate or other reducing agents) can provide an opportunity for injured or stressed microorganisms to recover.

Growth or decay (die-off) of microorganisms can also occur during sample collection, transport, and storage. In some cases, trace levels of toxic metals and/or organics may result in the inactivation of sensitive species, biasing the results. Dechlorination of chlorinated samples, use of EDTA to sequester trace metals, and the storage of samples at 4 °C, coupled with robust experimental design and quality assurance, can help to control some of these sources of error.

Examples of sampling locations include untreated wastewater, primary effluent, secondary effluent, filtered effluent, disinfected effluent, storage tanks, and/or distribution systems. Within a given sampling scenario, several factors influence the degree to which a grab sample can be used to represent the bulk liquid. Some water reclamation facilities include grit removal capability and equalization basins upstream of primary or secondary treatment, while in other facilities the water flows directly into the biological treatment unit. The presence of particulate matter and the degree of mixing and equalization can impact sample characteristics. In some cases, sampling ports are available, whereas in other cases it is necessary to obtain samples using pumps or bailing devices. It is important to ensure that samples reflect the characteristics of the sampling location by establishing rigorous protocols that include systematic flushing of sample ports prior to sampling, avoidance of contamination from sampling equipment, avoidance of cross-contamination, detailed record-keeping, etc.

2.3.1 Impacts of mixing on sample characteristics

Choices made about where and when to sample can influence the types and accuracy of conclusions that can be drawn from a microbial testing program. In a practical sense, because of the expense of sampling, it is important that samples are representative of "typical" conditions. The degree of mixing and potential short-circuiting at or near a sampling point can impact the measured concentration of microorganisms and therefore introduce a bias in interpreting the study results. In reality, every potential point of sample withdrawal reflects the specific hydrodynamic conditions that are prevalent during the sample event. Some of the tools that can be used to quantify temporal and spatial variations within treatment facilities include tracer studies, hydrodynamic models, and online monitoring. While online monitoring tools are not available for pathogen monitoring, online data for other parameters (e.g., pH, turbidity, chlorine residual, flowrates) can be used to qualitatively assess the treatment process variability. In general, collection of multiple samples can provide more insight into process variability; however, increased sampling requires additional funds and entails more complex statistical analyses (like the mixed-model analysis of variance [ANOVA]; Littell et al., 1996, Pinheiro and Bates, 2000) than will be covered in this document.

2.3.2 Timing of sample collection for process analysis

Choices made about when samples are collected can also affect the conclusions one can draw. In cases where the goal of the sampling program is to evaluate process performance or troubleshoot a treatment system, it is usually important to collect samples upstream and downstream of the treatment unit under investigation. Because flowrates and wastewater characteristics can vary over the course of a day, the extent to which microbial removal or inactivation occurs may also vary because of differences in loading rates, hydraulic detention times, temperature, mixing, and other factors. It is important that the time of day, the weather, the hydraulic and organic loading, the physical size of the service area and collection system, the type of collection system (sanitary or combined), and the instantaneous flowrate can influence the microbial composition. For example, peak flow conditions deliver higher volumes of water but reflect microbial inputs different from those reflected by low-flow conditions. The size of the service area and the amount of time that wastewater spends in the collection system also influence the microbial loading to a treatment facility. The loading rate affects the concentration of microorganisms; the flowrate affects the travel time between the wastewater source and the treatment facility. Flowrates also affect the hydraulic retention time of individual treatment units. For example, if there is an eightfold difference in flowrates between low flow and peak flow, then a disinfection basin that is designed for a 15-min contact time under peak flow conditions could provide a 2-h contact time under low-flow conditions. The longer contact time can result in increased inactivation of microorganisms compared to peak flow conditions. These issues need to be taken into consideration in determining the appropriate conditions for sample collection (location; time of day; day of week; and local situations that may impact loading such as sporting events, holidays, tourism, weather, etc.)

If a sampling program is designed to evaluate the effectiveness of an individual treatment unit or of the entire treatment facility, then samples need to be collected upstream and downstream (influent and effluent) of the components of the treatment facility that are to be investigated. For this situation, the timing of the sampling should reflect the hydraulic detention time where practical. For example, if it takes an average of 6 h for water to travel through a biological treatment unit, one would sample the inflow about 6 h before sampling the outflow, because the inflow concentration of microorganisms fluctuates over time. Alternatively, tracer tests could be conducted in parallel with the sample collection to estimate the hydraulic characteristics of the treatment unit. One would also want to begin analyzing the inflow and outflow samples right after obtaining each, to minimize artifacts caused by retaining living organisms in the samples for different periods of time (growth, die-off, predation, etc.). However, in practice, these "ideal" analyses are not always possible because of staff availability or other constraints. Such practical limitations may force one to collect inflow and outflow samples at about the same time. In any case, the statistical analyses reflect the conditions prevalent during sample collections and may not allow for differentiation between the relative importance of fluctuations over time versus differences due to treatment.

2.4 WHAT FACTORS INFLUENCE THE VOLUME OF SAMPLE NEEDED FOR EACH ANALYTE?

The volume of sample collected is directly related to the concentration of analyte that can be detected. In general terms, the limit of detection (LOD) represents the lowest average concentration (number per volume) of microorganisms that can be detected (quantified or result in a positive test) in an individual assay. In general, processing larger sample volumes increases the likelihood of detecting the target microorganism, as long as the intrinsic sensitivity of the assay is unaffected by sample volume. The minimum number of organisms that can be detected or grow in culture-based assays is one microorganism. If a 100-mL volume is sampled, the limit of detection for the assay is 1 per 100 mL. If the number is reported based on the growth of bacterial colonies, then the unit is CFU and the result would be reported as 1 CFU/100 mL or 0.01 CFU/mL. Similarly, if viruses are being measured, then the unit of measurement is PFU. In the case where a 100-mL sample volume yields zero colony, the result of this assay is correctly expressed as <1 CFU/100 mL. The reason is simple: if the source of our sample actually contains coliforms at a concentration of about one organism per every 500 mL, most 100-mL samples would not contain any bacteria, but they are in fact present and the one thing we can say with certainty is that there are fewer than 0.01 CFU/mL. Similarly, a sample volume of 300 mL may yield one colony. The detection limit in this case is 1 CFU/300 mL (0.003 CFU/mL). This detection limit is more sensitive than the 0.01-CFU/mL limit; the importance to study design is that one is more likely to detect low levels of microorganisms when the detection limit is lower. Large sample volumes have their own set of constraints. In many cases, samples are processed using filtration to separate the microorganisms from the water (or wastewater). Particulate matter in the sample, such as microbial flocs and organic or inorganic debris, can interfere with filtration by clogging the filter pores and limiting the amount of sample that can be processed. In addition, there is an upper limit to the number of cells that can be quantified using cell culture because of growth requirements. Thus, if too large a sample volume is processed, the result may be too numerous to count (TNTC). Other limits on sample volume relate to sample handling. Because microbial growth and die-off can occur after sample collection, it is important to minimize the time interval between sample collection and sample processing. In general, the larger the sample volume, the longer it will take to process the sample. In addition, transport of large volumes of water can be expensive.

Detection limits should also be relevant from a public health perspective. Because the minimum infective dose of pathogens can vary over several orders of magnitude, it is important that the detection limits are viewed from a health risk perspective. The health-relevant concentration (and detection limit) depends on considering the minimum infective dose and relative rates of growth and die-off associated with the pathogen in the given environment in conjunction with the potential for the public to be exposed to the reclaimed water, an estimate of the type of exposure (inhalation, ingestion, dermal contact, food, etc.), and the approximate quantity (volume) of reclaimed water to which the public may be exposed. Health-relevant concentrations may be above or below method detection limits, depending on the pathogen characteristics, prevalence, persistence, and growth requirements.

An example of the relationship between the concentration of microorganisms and the optimum volume of sample needed is shown in Figure 2-2. The optimum range of sample volumes is shown in the unshaded portion of the graph. If too small a volume is used, it may not be possible to detect microorganisms (lower triangle). Conversely, if too large a volume is used (upper triangle), accurate quantification may not be possible, particularly if cell culture methods are used.

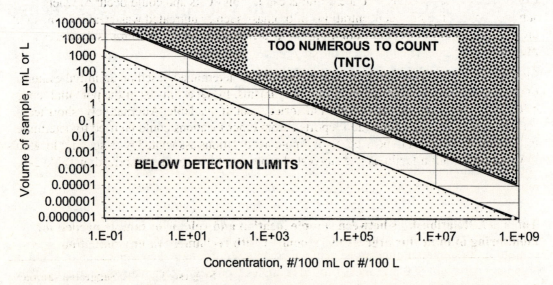

Figure 2-2. Volume of sample needed to detect and quantify microorganisms in relation to the actual concentration of microorganisms in a sample.

As shown in Figure 2-2, for concentrations of microorganisms below 0.1/100 mL (1/L), sample volumes in excess of 1 L are needed to detect the target analyte. Processing large volumes of samples can also introduce complications due to the potential for clogging of filters, overgrowth of nontarget microorganisms, or concentrations of target microorganisms that are TNTC. Thus, processing a sample volume that is too large may provide qualitative or categorical information on presence or absence, but quantification may be inaccurate or impossible. It is important that, in reality, not all microorganisms in a sample are captured through filtration or other concentration techniques. Inefficient capture can result from factors such as membrane filters with a pore size larger than some of the organisms (e.g., viruses) and filters for viruses or protozoa that do not retain all organisms. In other cases, the filters are effective at capturing the organisms, but inefficiencies exist in eluting or removing the organisms from the filter for subsequent testing (culture or molecular methods). Thus, the variability in methodology and efficiency must also be considered in selecting an appropriate sample size.

In general, the concentration of microorganisms in samples derived from reclaimed water facilities decreases with each treatment step. For monitoring protozoan pathogens (including *Giardia and Cryptosporidium*), high-volume filters have been developed for a sample concentration (e.g., USEPA Method 1623 [*Method 1623*, 1999]). For monitoring untreated wastewater, where concentrations may be in the range of 100-200 (oo)cysts/100L, sample volumes of 0.5 to 1 L may be adequate. However, since exposure to low numbers of oocysts can result in infection (DuPont et al, 1995), it is important that an adequate volume of water is processed to detect concentrations of a few oocysts in 100 L. Typically, to monitor reclaimed water, a minimum of 50 L must be processed. In addition to being time-consuming, processing such large volumes of reclaimed water may clog the filtration system because of the residual particulate matter (turbidity), colloids, and high-molecular-size organics. The consequences of filter clogging include lower net volumes of sample, higher detection limits, and potentially incomplete recovery of microorganisms. When one is using PCR methods, detection limits may be impacted by inhibitory compounds present in the reclaimed water. Conversely, excessive DNA concentrations can inhibit PCRs and could occur a) when samples containing high concentrations of biomass such as untreated wastewater or biosolids are analyzed or b) when other inhibitory molecules such as metals or humic substances are present.

Ideally, the appropriate volume of sample should be determined based on an initial estimate of the approximate concentration of microorganisms. Initial estimates of the potential range of concentrations can be derived from historical data, if available. Alternatively, short-term screening tests and/or information reported from other locations could be used to determine initial sample volumes to be used. Sampling considerations associated with specific locations in water reclamation facilities and suggested sample volumes are summarized in Table 2-7.

Table 2-7. Relationships between sample location and volume of sample needed for monitoring of indicator organisms associated with reclaimed water production

Sample location	Volume constraints	Potential sample-handling issues	Suggested detection limit for indicators[c]	Suggested sample volume for indicators
Untreated wastewater	Highly heterogeneous; solids may clog filters; use of several replicates is essential	Association of microorganism with particles	10^3–10^4/100 mL	Collect 100 mL and analyze a series of dilutions
Biological treatment	Lower concentrations of microorganisms; need higher volumes than influents	Association of microorganisms with particles	10–10^2/100 mL	Collect 500 mL and analyze a series of dilutions
Filtration	Low concentrations of indicators and pathogens; use higher sample volumes and multiple replicates	Need to assess resuscitation of "stressed" microorganisms	0.1/100 mL	Collect 1 L and analyze 10, 100, and 500 mL and a series of dilutions
Disinfection; distribution systems	Low concentrations of indicators and pathogens; use higher sample volumes and multiple replicates	Need to assess resuscitation of "stressed" microorganisms	0.01/100 mL	Collect 3 L and analyze as high a volume as practicable.

[a]Different indicator concentrations in influent vary widely; therefore, the range of sample volumes tested to obtain a given LOD will vary depending upon the organism.

2.5 WHAT IS THE APPROPRIATE NUMBER OF SAMPLES THAT SHOULD BE COLLECTED TO MEET THE GOALS OF THE STUDY?

A critical aspect of program design is to ensure that enough information is generated through sample collection to allow for statistical analysis. In general terms, the minimum number of samples that should be collected from a specific sampling site depends on the overall sampling goal(s). If we are interested in comparing data from different sets of data (before and after treatment, indicators and pathogens, different analytes, impacts of storage and distribution, impacts of loading rates, etc.), then these comparisons are statistically possible only if the degree of difference among the data sets is greater than the variation within each data set. Otherwise, if the variability of a given set of data is greater than the expected difference between (or among) groups of data, then it is difficult to discern a statistically significant difference.

For a given situation, the minimum number of samples required depends on two important factors:

- the degree of variability associated with a given microbiological measurement and
- the magnitude of the difference among different groups of samples.

In planning studies, it is important to have preliminary data to provide insight into the degree of variability associated with a specific parameter. It is also important to define the degree of difference that is important. In analysis of existing data sets, the variability of a set of data can be calculated and used to estimate the statistical **power** of the study. This topic is covered in more detail later in the document.

2.6 SOURCES OF ERROR DUE TO SAMPLING OR ANALYSIS

The potential for errors to occur should be considered early in the design of sampling programs for the microbiological analysis of reclaimed water. Quantification of microorganisms in reclaimed water involves several sequential steps, including sampling, sample processing (dilution and/or filtration and/or staining), and enumeration or detection. At each step, there is potential for introduction of error. A summary of potential sources of error due to sampling or processing of samples is given in Table 2-8.

A robust quality assurance program that includes replicate samples can help to overcome some of the potential sources of error outlined in Table 2-8. In general, analysis of multiple samples will result in a more robust set of data and help to identify potential sources of error. At a minimum, duplicate samples should be run on at least 10% of samples and at least one sample per group of samples (*Standard Methods*, 2005). It should be noted that the method of collecting replicates can influence interpretation of results. True replicates consist of discrete samples collected independently at the site of sampling. Pseudo-replicates consist of running parallel tests from a single sample, which measures only the precision of the analytical method. As a result, true replicates are measures of sample variability, while pseudo-replicates are measures of analytical variability. A comparison of true and pseudo-replicates is shown in Figure 2-3. A combination of true and pseudo-replicates provides information on analytical and sample variability.

Table 2-8. Sources of error associated with enumeration of microorganisms from reclaimed water

Step	Potential source of error	Corrective measure
Sample collection		
	Variable water quality at sampling location.	Take multiple samples (replicates) and process each independently.
	Treatment process modifications or process upset prior to or during sampling.	Collect detailed information on operations to help interpret anomalies in data; conduct repeat sampling after system has stabilized.
	Inability to obtain representative sample of reclaimed water from the specific sampling location.	If sample location is accessible, use pump or bailer to try to obtain sample; if sample is from a closed system (e.g., pipe, dedicated sample tap), flush the system with enough velocity and volume to purge stagnant water prior to sample collection.
Sample processing		
Dilution	Inaccurate volume of sample or dilution media.	Process multiple samples (resample and process with a wider range of dilutions or volumes).
Filtration	Clogging of filter; loss of microorganisms within filtration device.	Make sure filtration volume is appropriate for characteristics of sample; use multiple filters in parallel.
Enumeration		
	Presence of particulate matter that shields microorganisms.	Use sonication or mixing to disperse microorganisms before processing of sample.
	Sample contains residual disinfectant that acts to inactivate microorganisms during sample transport.	Add thiosulfate to samples from water that has been chlorinated (5 mg of thiosulfate/mg of chlorine residual), chloraminated, ozonated, or treated with chlorine dioxide or other oxidants (permanganate and hydrogen peroxide).
	Samples contain VBNC organisms.	Use resuscitation techniques; incubate samples for longer periods (use a gradual ramping of temperature from ambient to incubation temperature as to reduce "shocking" cells).
Data interpretation		
	Results are reported as TNTC.	Determine if results can be interpreted as categorical data instead of as quantitative data. Use statistical methods to develop estimates of concentration; collect repeat samples with smaller volumes (or filterable volumes with a higher dilution factor) and more replicates.
	Results are below detection limits.	Determine if results can be interpreted as categorical data instead of quantitative data. Use statistical methods to develop estimates of concentration; collect repeat samples with larger volumes.

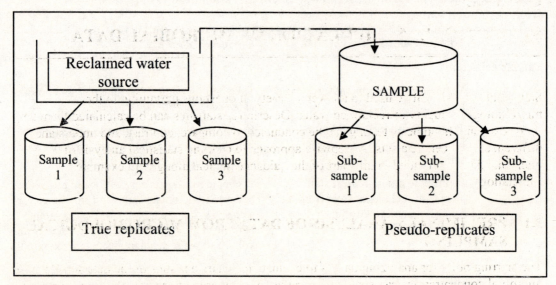

Figure 2-3. Comparison of true replicates and pseudo-replicates for microbiological characterization of reclaimed water.

CHAPTER 3

STATISTICAL MANIPULATION OF MICROBIAL DATA

Statistical analysis can be used to answer a variety of questions pertaining to the microbiological quality of reclaimed water. Descriptive statistics can be calculated from any set of data, and hypothesis testing can be conducted to compare, contrast, and understand differences between data sets. A stepwise approach to tackling statistical analysis of microbial data is provided in this part of the guidance manual along with example calculations.

3.1 PRELIMINARY ANALYSIS OF DATA FROM MICROBIOLOGICAL SAMPLING

The starting point for analyzing data is to evaluate the characteristics of the data set. Because microbial concentrations can span a wide range, it is important to review the data set prior to conducting statistical tests. Data can be represented in a table format or with a variety of graphical approaches. If the data are displayed directly, then the range of values can be compared. In some cases, it is more appropriate to use a log scale to display the data. The log scale allows for comparisons over a wider range of concentrations than does a linear scale.

3.1.1 Descriptive statistics

A review of the data should include a sense of the overall range of the values. The easiest way to do this is to graph the data. There are several computer packages that allow for fairly easy generation of frequency diagrams. Examples of histograms of data on arithmetic and logarithmic scales are shown in Figure 3-1. Probability plots of data are also shown in Figure 4. Note that the log-transformed data (Figure 3-1b) have a bell-shaped distribution, while the untransformed data (Figure 3-1a) have a highly skewed distribution. The importance of data distribution is discussed in the next section.

To describe a set of data, it is useful to present the number of data points, the range, a measure of central tendency (mean, median, and/or mode), and the standard deviation. Other descriptive statistics include the skewness and kurtosis, which give an indication of the symmetry (or lack of symmetry) of a distribution.

CASE STUDY

BOX 4a: PRELIMINARY ANALYSIS OF DATA

In anticipation of a filter upgrade, a treatment facility is compiling baseline data on filter performance. Grab samples of the filter effluent have been collected once/month and analyzed for total coliforms, and the data are tabulated.

It is useful to graph the data to evaluate the overall characteristics of the data set. For statistical analyses, it is also important to characterize the distribution of the data.

Because microbiological data typically span several orders of magnitude, it is common practice to log-transform the data. This is done by taking the $\log_{10}$ of each observation. The data are tabulated in CASE STUDY BOX 4b.

CASE STUDY

BOX 4b: CONCENTRATIONS OF TOTAL COLIFORMS IN FILTER EFFLUENT

Results of monthly grab samples of filter effluent are shown in Table CS-4. The numerical value and the $\log_{10}$-transformed value are shown. The data can be graphed to evaluate the overall distribution. Statistical calculations such as mean, standard deviation, and other measures of central tendency and variability along with determining whether parametric tests can be applied are based on how the overall distribution of data can be modeled.

Month of sample collection from combined filter effluent	Total coliforms	
	CFU/100 mL	$\log_{10}$ of total coliforms
Jan	108,895	5
Feb	2117	3.3
Mar	1810	3.3
Apr	35,217	4.5
May	4.17	0.6
June	3678	3.6
July	75,645	4.9
Aug	37	1.6
Sept	0.4	−0.4
Oct	1269	3.1
Nov	127	2.1
Dec	1.2	0.1

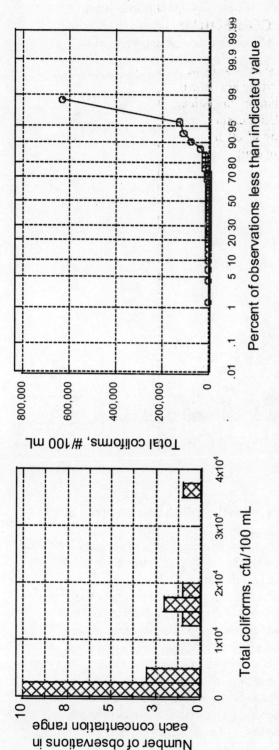

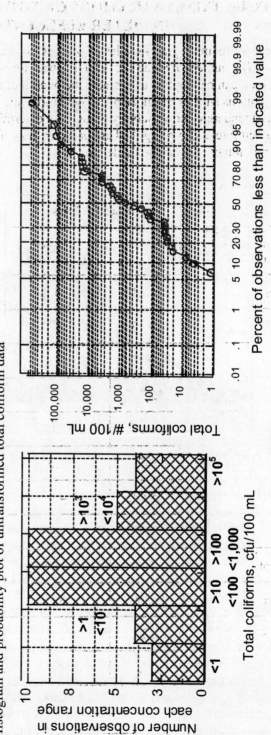

a. Histogram and probability plot of untransformed total coliform data

b. Histogram and probability plot of log$_{10}$-transformed total coliform data

Figure 3-1. Graphical representation of data set: a) histogram and probability plot of 35 observations, b) histogram and probability plot of log$_{10}$-transformed data (same 35 observations as shown in panel a).

25

3.1.2 Role of data distribution in statistical analyses

Many statistical tests are based on the assumption that data are "normally" distributed, which means that the range of datum values can be modeled as a bell-shaped distribution. When data are normally distributed, the majority of the data are grouped between two extremes and the central value is halfway between the highest value and lowest value, as shown in Figure 3-2. In the figure, "+/- 1s, 2s, and 3s" refers to 1, 2, and 3 standard deviations of data. The standard deviation is a measure of how closely the data are clustered around the mean in a set of data. When the datum values are very close to the mean, the bell-shaped curve is steep, the standard deviation is small, and the data have a higher degree of precision. Conversely, the distribution associated with datum values that are more distant from the mean is relatively flat, suggesting a higher standard deviation and lower precision. Statistical tests based on data that are normally distributed are referred to as "parametric" tests.

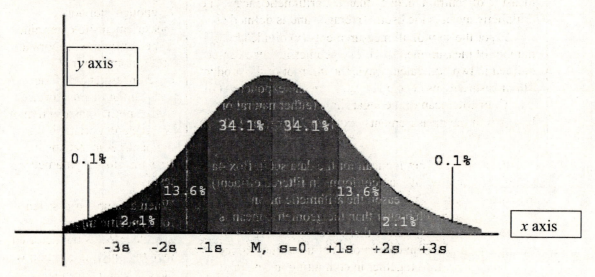

Figure 3-2. Generic example of a normal distribution curve.

An important and useful property of normal distributions is that the bell-shaped curve can be subdivided based on the relationship between the standard deviation and the mean. In general, 68% of the data are contained within 1 standard deviation from the mean, 95% of the data are within 2 standard deviations, and 99% of the data are accounted for by taking the mean value plus or minus 3 standard deviations. The ratio of the standard deviation to the mean is referred to as the coefficient of variation (CV = SD/Mean) and can be used to compare the variability among sets of data. For microbiological data, it is not uncommon to observe coefficients of variation ranging from 5 to 20%. Because microbiological concentrations can vary over several orders of magnitude, the assumption of a normal distribution is not always appropriate. In same cases, data can be mathematically transformed to mimic a normal distribution. A common type of transformation is a log transformation in which the logarithm of the concentration is used in statistical calculations instead of the actual concentration. In many cases, log-transformed data can be described by normal distributions allowing for the use of parametric statistical tests. Sometimes, the distribution of the data may be bimodal or trimodal—in other words, there are two, or even three, peaks in the frequency histogram of the data, rather than only one. In some cases, bi- or trimodal data can be subdivided into discrete groups of normally (or log-normally) distributed data. In other cases, log transformations of data do not result in normal distributions.

BOX 5: DESCRIPTIVE STATISTICS

The first step in describing a set of observations is to assess the type of distribution. A variety of tests can be used to assess the degree to which a distribution can be described as a normal distribution. The data shown in Table CS-5a are not normally distributed. However, the $\log_{10}$ transformation of the data can be represented by a normal distribution (see Figure 4).

When describing data that are log-normally distributed, it is more appropriate to use the geometric mean as a measure of central tendency than the arithmetic mean. The arithmetic mean is the usual "average" and is defined as $1/n \; \Sigma x_i$ or the sum of all measurements (x) divided by the number of measurements (n). The geometric mean can be defined in two equivalent ways: the nth root of the product of n measurements $(x_1 x_2 x_3 \ldots x_n)^{(1/n)}$ or the exponential of the arithmetic mean of the logarithms (either natural or base 10) of the measurements ($\exp\{1/n \; \Sigma \ln x_i\}$ or $\{10^\wedge 1/n \; \Sigma \log_{10}[x_i]\}$).

The mean and geometric mean for the data set in Box 4a (monthly sampling of total coliforms in filtered effluent) are shown below. The reason the arithmetic mean (average) is so much higher than the geometric mean is that there were a few values that were several orders of magnitude higher than the majority of the values. Since the values are added together in computing the average, the high values mask the presence of the lower values, skewing the result. The log transformation tempers this impact.

Parameter	Example calculation (values from Box 4a, $n = 12$)	Mean or geometric mean concentration, CFU/100 mL
Mean (average)	$228{,}801 \big/ 12$	19,067
Geometric mean	$10^{32/12} = 10^{2.64}$ or $\exp\!\left(73/12\right) = \exp(6.1)$	437

3.1.3 Approaches for handling measurements that are below detection limits

In samples where the concentrations of microorganisms are low, the detection limits may introduce an artificial lower boundary to the range of data, causing a distortion or truncation in the distribution. In other cases, there may not be enough individual measurements to determine if the distribution is normal or log-normal or has a different structure. If the data cannot be transformed to fit a normal distribution, it is more appropriate to use nonparametric statistical tests than to use parametric tests.

When a sample is assayed for a specific microbial analyte and it is not detected, then the only information we have is that it was not detected under the sampling and experimental conditions used for the test. We cannot conclude that the microorganism is not present, just that it was not detected.

The closer the detection limits are to 1 microbial unit (e.g., 1 CFU or one oocyst), the more confidence we have that the organism is not present in the volume sampled. In fact, nondetection of microbial pathogens is a positive outcome of many monitoring programs, provided the detection limits are relevant to human health outcomes or to the overall objectives of the sampling program.

Measurements that are below the limit of detection can result from a variety of circumstances including

- inadequate sample volume (see Figure 2-2),
- physical or chemical interference, including microbial association with particles, and/or
- absence of the microbial analyte.

While careful planning of testing protocols can help control the number of measurements that are below detection limits, it is important to develop sound approaches for interpreting this type of data and maximizing the amount of information that can be derived from a sampling program. The starting point is to assess the question that is to be answered and determine how nondetected observations may impact the decision. Examples of how nondetected observations can impact conclusions for some of the major applications of microbial monitoring of reclaimed water are summarized in Table 3-1.

Table 3-1. Impact of nondetected observations of microbial analytes on sampling programs associated with evaluation of reclaimed water

Sampling program goal	Impact of nondetected observations
Compliance monitoring	Regulatory requirements are either based on a fixed numerical limit or presence or absence. As long as the detection limit is below the numerical limit, nondetected observations can be used to demonstrate compliance.
Assessment of microbial reduction through treatment	Typically reduction of microorganisms is calculated as log reduction, which is defined as **$\log_{10}$ (concentration before treatment) minus $\log_{10}$ (concentration after treatment).** For this calculation, a numerical value (>0) is needed for both parameters: the influent (concentration before treatment) and effluent (concentration after treatment). If the concentration after treatment is nondetectable, it is not possible to calculate a "true" log reduction, since the log of 0 is not defined and substituting a value (or zero) can bias the estimate. Similarly, if the initial value is nondetected and the final value is detected (because of changes in the volume sampled or water quality), log reduction calculations may not be realistic (unless there was a true increase in concentration following treatment). In many cases, substituting "zero" values for nondetects (by eliminating the term from the calculation) can lead to an overestimation of the log reduction, particularly when the LOD is large.
Correlation or comparisons among water quality, process, and/or microbial parameters	If there is a need to relate microbial concentrations to other parameters such as BOD, TSS, turbidity, nutrients, or other chemical or microbial analytes, nondetected observations cannot be used to develop quantitative relationships or empirical models. It is possible to use nondetected observations to develop categorical relationships.
Treatment process optimization and troubleshooting	If a treatment system is to be optimized to improve removal or inactivation of microorganisms, then reliance on observations below detection limits may yield false conclusions.

CASE STUDY

BOX 6: USE OF TOBIT REGRESSION TO ESTIMATE MEAN CONCENTRATIONS OF *CRYPTOSPORIDIUM* IN DISINFECTED EFFLUENT

Twelve disinfected effluent samples have been assayed for *Cryptosporidium* over the course of a year. The data and the detection limit for each measurement are shown below. About 50% (6/12) of the values are below detection limits. The mean concentration is estimated by Tobit regression and compared to other approaches.

Month	Concentration, oocysts/100 L	Detection limit, oocysts/100 L
Jan	<4	4
Feb	<9	9
Mar	29.4	4
April	<9	9
May	5.5	4
June	<8	8
July	12.2	4
Aug	<8	8
Sept	43.7	4
Oct	<6	6
Nov	9.1	4
Dec	16	4

Treatment of nondetected observations	Mean concentration, oocysts/100 L	Standard error of the mean	Confidence interval
Tobit regression	11.9	0.29	11.4–12.5
Substitutions:			
Zero	9.7	4.05	1.7–17.6
LOD	13.3	3.37	6.7–19.9
½ LOD	11.5	3.69	4.3–18.7
Omit nondetects	19.5	5.91	7.7–30.9

For this set of observations, the Tobit regression provides the lowest standard error and the narrowest confidence interval. The substitution methods either underpredict (zero substitution) or overpredict (LOD substitution) the mean concentration. Omitting the nondetected values reduces the number of data points and overpredicts the mean

From a statistical perspective, it is important that there is not one universal approach to address the use of data that are below detection limits that can be applied to all situations. In all cases, a systematic approach should be developed that is relevant to the goals of the sampling program. Some common approaches for dealing with nondetect observations in statistical calculations include

- ignoring the observation (eliminating it from the data set)
- assigning a value to the nondetected observation.

Typical approaches for assigning value include "zero," the detection limit, or half of the detection limit. Unfortunately, all of these approaches can introduce bias into the statistical calculation. Sometimes the bias is small and inconsequential; however, in other cases, a false conclusion may be drawn from this approach. For example, if there are only two data points below the LOD and 30 above it, the effect of ignoring those data points (or using the other three arbitrary substitution approaches) may be small. As the proportion of points below the LOD grows larger, though, the bias becomes increasingly severe—to the point where the "answers" from the statistics may be misleading. It is important that *caution should be exercised in interpreting nondetect data. Even if there are only a few measurements below the LOD, it is not possible to tell how severely statistical conclusions have been biased without more information.*

Statistical methods based on *maximum likelihood* approaches are particularly useful in handling censored data. Maximum likelihood methods involve two steps. First, one selects a statistical model and assumes that it is a correct description of the data—the same assumption is made in all statistical methods. Second, one finds the model parameters that would make it most probable that one would observe the data. This is done by numerically maximizing a *likelihood function*. The details of the likelihood function depend on the particular model, because it incorporates all the probability terms in the underlying statistical model. A comprehensive introduction to likelihood methods is given in Edwards (1992).

Fortunately, there is a method that can solve this problem, called censored regression or Tobit regression, which is a likelihood method. A number of statistical packages can estimate a Tobit regression, including R (http://cran.r-project.org), SAS (http://www.sas.com/ technologies/analytics/statistics/stat/index.html), and Stata (http://www.stata.com). The term "censored" refers to data points for which only a bound is known—in this case, we know that the values of certain data points are somewhere between zero and the LOD. Censored regression models can use the information that some data points have known values, while others are somewhere between zero and the LOD, to jointly estimate a model for the entire data set. This approach has been successfully applied to a number of situations (Lorimer and Kiermeier, 2007).

A comparison of the advantages and disadvantages of typical strategies for dealing with nondetected observations is given in Table 3-2. The optimum approach depends on the answers to the questions below:

- What question is to be answered?

- What percentage of the observations represent detected values versus values that are below the detection limits?

- What are the consequences of underestimating (using zero) or overestimating (using LOD) the true value?

The USEPA has developed statistical guidelines for practitioners (*Data*, 2006) that are applicable to microbiological testing of reclaimed water. The approaches for handling values below detection limits suggested in the USEPA guidelines are based on the frequency of nondetects within a given data set. A summary of the recommended approach and potential problems associated with applying this approach to microbiological characterization of reclaimed water is given in Table 3-3. As shown, there are several approaches presented for analysis of data sets where fewer than 15% of the values are below detection limits (substitution of a value or of zero and maximum likelihood estimate). The approach selected can influence the interpretation of the data, depending on the magnitude of the values, the detection limits, and the number of observations. For example, if datum values are fairly close to zero (0.001/mL), then the error associated with assuming the nondetected data are 0.001 or zero is likely to be less significant than if the detection limit is much greater than zero (100/mL) and if the measured values are close to the detection limits. In practice, dealing with nondetected values is problematic and there is no single calculation approach that will work under all circumstances. It is important to evaluate the strengths and weaknesses of alternative approaches to determine the optimum calculation approach for a given set of data.

Table 3-2. Strategies for statistical analyses of data sets that include measurements below detection limits

Strategy	Advantages	Disadvantages	Applications
Discard nondetects	Simple	Distorts the data set by eliminating observations; Does not allow for interpretation of conditions that result in nondetected values (e.g., removal and/or inactivation of microorganisms).	Screening of data for correlations between/among microbial analytes and other water quality or process data; Evaluation of data distributions.
Treat as zero	Simple, avoids overestimating	Could underestimate significantly, especially when LOD is >> 0; Could skew data set.	Estimate of sample statistics (mean, median, skewness, standard deviation, etc.).
Treat as ½ LOD	Simple, compromise	Could overestimate significantly depending on relationship of LOD to measured values and the percentage of nondetected observations; Could skew data set.	Calculation of log reduction associated with a treatment system; Estimate of sample statistics (mean, median, skewness, standard deviation, etc.).
Treat as LOD	Simple, conservative (avoids underestimating)	Could overestimate significantly depending on relationship of LOD to measured values and the percentage of nondetected observations; Could skew data set.	Calculation of log reduction associated with a treatment system; Estimate of sample statistics (mean, median, skewness, standard deviation, etc.).
Use censored (Tobit) regression models	Includes all data in statistical analysis, including nondetects	Requires that a sufficient portion of the data points be quantified.	All applications: Data distributions, Log reduction, Sample statistics, etc.

Table 3-3. Guidelines for interpreting nondetect data based on the percentage of samples below detection limits (adapted from *Data*, 2006)

Observations below detection limits, %	Approach	Assumptions	Comments
<15%	Substitute a value for the nondetect samples.	Assumes detection limit is low relative to observed values. Values substituted include detection limit, half of the detection limit, random number, or zero.	The potential error depends on the magnitude of the detection limit. Use of detection limit will yield a conservative estimate of log reduction.
<15%	Adjust mean and standard deviation by assuming nondetects are zero (Aitchison's Method).	Assumes detected values are normally distributed and a proportion of values are zero. Typically, microbiological data are assumed to be log-normally distributed.	This method may result in overestimation of log_{10} reduction, particularly if the detection limit > 10.
<20%	Maximum likelihood estimation to estimate mean and variance (Cohen's Method).	Assumes all data are normally (or log-normally) distributed and the detection limit is the same for all observations. Need to have >20 observations to yield consistent results.	The statistical power decreases as the proportion of observations below the detection limit increases.
>50%	Test of proportions.	At least 10% detected.	Categorical data (presence or absence) may yield more accurate results.

3.2 EVALUATION OF LOG REDUCTION

A typical approach for evaluating the effectiveness of individual treatment processes is to compare the difference between the concentration of microorganisms that enters a treatment system and the concentration exiting the system. Treatment systems used to produce reclaimed water are frequently modeled as first-order (linear) reactions. Therefore, an often-used approach to describe the reduction of microorganisms is to compare the concentration of microorganisms at each stage of treatment on a logarithmic scale.

Log reduction is a term that is widely used to describe the change in concentration between two sampling locations, such as the influent and effluent to a treatment unit or a treatment facility. Log reduction is calculated as

$\log_{10}$ reduction = $\log_{10}$ (initial concentration) - $\log_{10}$ (final concentration)

This can also be expressed as:

$$Log_{10} reduction = Log_{10}\left(\frac{InitialConcentration}{FinalConcentration}\right)$$

Alternatively, treatment systems are described in terms of removal efficiency where

$$\%removal = 100\%\left(\frac{InitialConcentration - FinalConcentration}{InitialConcentration}\right)$$

90% removal corresponds to 1 $\log_{10}$ reduction, 99% removal corresponds to 2 $\log_{10}$ reductions, 99.9% removal corresponds to 3 $\log_{10}$ reductions, etc.

When actual values exist for the concentrations of microorganisms entering and leaving a treatment system, calculation of $\log_{10}$ reduction is fairly straightforward. If either the influent or effluent concentrations are below the method detection limits, then a systematic approach should be applied to interpret $\log_{10}$ reduction. It should be noted that the use of "zero" is mathematically incorrect since the $\log_{10}$ of "zero" is undefined, as is a "zero" in the denominator. Also, by definition, the values being compared should be "paired" (see section 3.3); that is, that the initial and final concentrations should reflect a single sampling event. The important factors are

- the magnitude of the detection limit,
- the relationship between the concentration of the upstream process and the detection limit,
- the total number of detected samples and statistical power, and
- the percentage of samples that are below detection limits.

To some degree, methods for handling nondetect data for calculating $\log_{10}$ reduction depend on how the information will be used and how many observations are in each set of data.

3.3 HYPOTHESIS TESTING

In general, statistical testing is based on defining one or more hypotheses and testing to see if the data support the assumptions. The overall approach is to contrast the hypotheses to be examined with a "null" hypothesis that states that the factors considered in the other hypotheses have no effect. For example, suppose we were interested in evaluating the efficacy of medium filtration for removal of *Cryptosporidium*. The null hypothesis is that concentrations in postfiltration samples will not differ from the unfiltered water.

When microbiological data are collected, the goal is to provide enough information to reject a specific null hypothesis if it is actually wrong. For example, if one is interested in evaluating the effects of storage on the concentration of enteroviruses in water, the null hypothesis would be the following: there is no significant difference in the concentration of enteroviruses before storage and after storage. If the null hypothesis is rejected, then one can conclude that the concentration of pathogens changes (increases or decreases) during storage of reclaimed water.

One may also be interested in determining if there has been a significant change in the concentration of microorganisms in reclaimed water due to treatment. In this case, statistical tests are applied to verify the hypothesis that there is a difference between the means of two or more groups of data and the null hypothesis would be that there is no difference between the means of the data sets. If we are looking for evidence of a difference—say, between the concentrations of bacteria in the influent at two different treatment plants, but are not concerned about which set of concentrations is higher, then the statistical tests are classified as nondirectional or "two-tailed." On the other hand, if we are considering whether one type of treatment removes more *Cryptosporidium* oocysts than another, the tests are directional or "one-tailed." The "tails" represent the extreme (high and low) values of a normal probability distribution (Figure 3-2). The right tail represents unusually high values, and the left tail represents unusually low values. Because they are unusually high or low values, the probability of getting values in one of the tails by chance alone is very small. If a one-tailed (directional) test is selected, then, in some cases, fewer data may be needed to obtain statistical significance than when one is using two-tailed (nondirectional) statistical tests. In other cases, there are minor differences in statistical requirements.

3.3.1   When are paired data advantageous?

Another important consideration in hypothesis testing is consideration of data as "paired" versus "unpaired" observations. Pairing or matching data by some attribute, for example, time or location, tends to reduce the variability between observations and therefore allows more sensitive detection of a true difference between treatments. If two quantities are being compared, it may be important that they are measured at the same time and under the same circumstances (paired). A simple example is in a study of the efficacy of a particular treatment—say, chlorination. If the treatment unit is completely mixed, then it is assumed that the concentration leaving the unit is at steady state. However, for plug-flow treatment systems or water distribution systems, it is typically assumed that water moves through the system as a "slug", and therefore, where practicable, sampling should account for the hydraulic residence time

Similarly, if one wanted to know the relationship between turbidity and microbial concentrations or between concentrations of pathogens and indicators, one would need to measure these quantities from the same sample. In this case, the paired data can be used to evaluate the degree to which the two independent measurements can be correlated.

Whether paired data are collected depends on the hypotheses being studied. If the specific goal of the sampling is not dependent on paired data, then the design of the sampling program should reflect this. For example, if microbiological data from eight different treatment facilities are being compared to determine if there are differences in the concentrations of bacteria after filtration, unpaired data can be used. Since the samples are derived from different sources, there is no reason that the samples need to be collected on the same day. However, one would certainly want the data to reflect similar conditions. For example sampling reclaimed water facilities before, during, or after major storm events may impact flowrates and loading rates and hydraulic retention times (depending on the service area and the collection system). To the extent practicable, it is important that the sample locations, volumes collected, and other conditions are comparable and consistent.

Examples of hypotheses related to microbiological testing of reclaimed water are listed below, and the types of data that would be required to test each hypothesis are given in Table 3-4.

1. The reduction (removal and inactivation) of indicator organisms through wastewater treatment is similar to the reduction of pathogens.

2. An increase of the contact time in a chlorine contact basin results in improved inactivation of pathogens.

3. Addition of ammonia to a chlorine contact basin (to form chloramines) does not reduce disinfection effectiveness.

4. The concentration of pathogens that persist through wastewater reclamation is related to water quality parameters such as BOD, TSS, TDS, and nutrient levels.

Table 3-4. Examples of the type of data needed to test specific hypotheses related to the microbiological characteristics of reclaimed water

Goal of test	Type of data needed
Comparison of indicators and pathogens	Parallel samples of concentrations of indicators and pathogens from the same sample site (paired data)
Comparison of contact time and pathogens	Concentrations of pathogens as a function of contact time
Comparison of chemical addition, indicators, pathogens	Parallel samples of concentrations of indicators and pathogens under different conditions of chemical addition (paired data)
Comparison of water quality data and pathogen data	Historical data of pathogen concentrations and water quality data from specific treatment units or over a specific time frame (paired data)

3.3.2   What is the significance of significance?

Before developing an approach for hypothesis testing, it is important to understand what is meant by the term "significant." This seemingly simple word often causes confusion. There are two aspects to this confusion. First, many people are unclear on the statistical notion of significance and what it means. Informally, the question can be posed as:

How likely is it that the difference between two sets of observations would occur by chance alone?

If it is very likely that the result occurred by random chance, the test results are considered "not significant." It is important that the cutoff point for "very likely" is arbitrary. While 5% is widely used, there is no absolute justification for setting the cutoff point at 5% as opposed to 1% or 10%. Indeed, R. A. Fisher, one of the founders of modern statistics, suggested that investigators should select their cutoff point carefully for each study, based on everything they know about the system—a sharp contrast with the current somewhat arbitrary practice of applying 5% as an absolute cutoff point for all studies.

The second kind of confusion is that "statistical significance" has a different meaning from "public health significance." For example, consider a reclaimed water facility that has conducted pilot testing of alternative filters to compare removal of total coliforms by a new technology to the performance of the existing facility. We are interested in figuring out if the new technology is "better" than the existing treatment facility. Because of inherent variations in coliform concentrations, it is unlikely that the mean concentrations of total coliforms from the pilot test and the full-scale plant operations would be identical. Therefore, it is important to figure out if the difference between the removal efficiency of the two systems is *statistically significant*. It is important that even small or minor differences can be statistically significant, depending on the number of observations and the variability among the observations. Depending on the number of samples collected, it may be possible to conclude that a small difference in total coliform concentrations (2/100 mL) is *statistically significant*. But this conclusion means only that the difference is not likely to occur by chance alone, based on the cutoff level chosen for the study. While such a conclusion provides a basis for comparing treatment systems, it does not imply that the difference is of any public health significance. Will there be more illnesses if the treatment facility does not install the new process? That conclusion does not follow from a statistically significant different of two total coliforms per 100 mL. The reverse situation can also be true: a difference can have public health significance (in the sense of increased illness) but, because of limits on the sample size or detection limits, not be statistically significant. In addition, monitoring programs frequently rely on indicator organisms that may or may not be related to the etiologic agents of illness and there is a time lag between exposure and illness.

3.4 STATISTICAL SOURCES OF ERROR

While it is not possible to completely avoid errors in microbiological testing or statistical analyses, careful planning of each study coupled with a sound quality assurance program will help to control the effects of errors on data interpretation. Accuracy of data interpretation and the data's ultimate usefulness are limited by the quality, structure, and quantity of data (*Guidance*, 2006).

An important initial point for statistical analyses is selecting acceptable error rates that are appropriate for the question(s) at hand. The alpha level is often called the "level of significance" and defines one type of error (Type I), the probability of rejecting the null hypothesis when the null hypothesis is true. By convention, the null hypothesis states that two groups are NOT different or that there is NO relationship between variables. For example, if one is trying to determine whether the efficacy of disinfection for *Cryptosporidium* by chlorination versus ultraviolet radiation treatment differs, the null hypothesis could be framed as "the log reduction of *Cryptosporidium* after chlorination is not different from the log reduction after ultraviolet treatment." In this example, when one **rejects** the null hypothesis, it is because the treatments have been found to produce significantly different results. If the null hypothesis is rejected when it is in fact true, it means that one has found a statistically significant difference when one does not exist. In statistical analysis we tend to strongly guard against Type I error and typically choose an alpha level ranging between 0.01 and 0.1.

While the selection of the alpha level is somewhat arbitrary, it is fairly common to use an error rate of 0.05. An alpha level of 0.05 means that there is a 5% likelihood that the difference between two groups of data arises from random chance and there is a 95% probability that there are real differences (statistically significant) between (or among) the sets of data. Put another way, one would expect that if a statistical test shows a difference at the alpha level of 0.05, the results from 100 repetitions of the study would reveal that 95 out of the 100 trials were different, while there would be no apparent difference in the other five trials. The alpha level is also defined as the probability of occurrence of a Type I error.

There are two general types of statistical errors, as compared in Table 3-5. Type I errors can result from selecting an alpha level that is less stringent than is warranted by the variability of the data. In addition, the structure (distribution) of the data must be compatible with the type of statistical analysis that is conducted. Beta, or Type II, error is the probability of accepting the null hypothesis when it is false. The consequence of Type II error is that one would fail to find a difference when one really exists. Type II errors can result from trying to draw conclusions from an inadequate number of samples or from substituting a high LOD value for nondetect observations. Because of the complexities of microbial sampling, it is possible that several independent errors can occur within a single data set, such as inadequate sample numbers or volume, creating additional confusion in data interpretation. In designing studies or in retrospective analysis of data, it is important to be cognizant of the impacts of statistical errors on data interpretation.

3.5 RELATIONSHIPS BETWEEN THE NUMBER OF SAMPLES COLLECTED AND SAMPLING GOAL(S): POWER ANALYSIS

To draw statistically sound conclusions about the microbiological characteristics of reclaimed water it is important to have an adequate amount of information available. From the perspective of design of microbial sampling programs, the minimum number of samples that should be collected from a specific sampling site depends on the overall sampling goal(s). The key variables that affect the minimum number of samples required for a given situation are the degree of variability associated with a given microbiological measurement and the magnitude of the expected difference between different groups of samples. To compare different data sets, the degree of difference among the data sets needs to be greater than the variation within each data set. Otherwise, if the variability of a given set of data is greater than the expected difference between groups of data, then it is difficult to discern a significant difference.

In statistical terms, the power of a study relates to the ability of the data to refute the null hypothesis if it is actually false. Terminology related to power analysis is defined in Table 14 with respect to determining the appropriate number of samples needed to answer a given question. It is important that the alpha level and the standard deviation also impact statistical power. Higher values of alpha for a given set of data (number of samples) can yield more statistical power. Similarly, the lower the standard deviation (or coefficient of variation) associated with a given set of data, the more statistical power is associated with a given sample size.

Table 3-5. Comparison of statistical errors that can result from analysis of microbiological data

Type of error	Definition	Cause(s)	Impact on conclusion	Corrective measures
Type I error	An error resulting from concluding that the null hypothesis is false when it is true. Differences between the test results are due to chance, not "true" differences. Probability of Type I error is equal to alpha, the significance level.	Discrepancy between test sensitivity and significance level. Structure (distribution of data may not meet the underlying assumptions of the statistical tests (e.g., normal distribution).	Test results indicate that there is a difference when there isn't one: False positive.	Increase the sensitivity of the test by decreasing the significance level (alpha value). Validate data structure or transform data if appropriate.
Type II error	An error resulting from concluding that the null hypothesis is true when it is false. Probability of Type II error is beta. Power (probability of rejecting a false null hypothesis) is 1 – beta.	Inadequate power to discriminate differences among samples.	Test results indicate that there isn't a difference when there is one: False negative.	Increase the number of samples or sample volume collected to provide more statistical power.

Table 3-6. Key factors that impact the minimum number of samples needed to evaluate the microbiological quality of reclaimed water

Term	Definition	Application
Significance level (alpha)	An estimate of the degree to which the differences between groups of samples are not due to random chance. This is also referred to as the alpha value or Type I error and is the probability that the test statistic will fall in a critical region, resulting in rejection of the null hypothesis, when the hypothesis is actually true. Typical values for alpha range from 0.01 to 0.1, and most commonly 0.05 is used (but not specifically justified). Lower values of alpha suggest a higher degree of confidence (precision) in a statistical conclusion (e.g., accepting or rejecting the null hypothesis).	The significance level is a tool for comparing different groups of data or testing the validity of a "null" hypothesis. The number of samples needed to detect a given effect with a specified probability is related to the significance level (i.e., more samples are needed for lower values of alpha).
P	P is the outcome of statistical hypothesis testing and is the probability that the observed result occurred by chance. Results are considered to be statistically significant if P values are below a specified cutoff alpha value.	For a given alpha value, results are significant if $P <$ alpha (e.g., the familiar $P < 0.05$).
Sample size	The number of samples (sampling events) used to test the null hypothesis and answer the specific question posed by the study. An adequate sample size helps to yield reliable information, whereas an inadequate sample size makes it difficult to interpret negative information. If the sample size is too small, it is impossible to draw meaningful conclusions.	Samples need to be collected from each location relevant to the study question. Increasing the number of samples while holding the significance level (alpha) constant usually results in higher statistical power.
Effect size	The difference in average measurements between groups being studied, such as samples from different sites or samples from a single site over different time. The effect size can also be reported as percent difference.	The difference in concentration between the test groups and the analytical variability affect the number of samples needed and the overall statistical power.
Coefficient of variation	The ratio of the standard deviation to the mean for a given set of data.	Comparing data sets.
Power	The probability of obtaining a significant P if a true difference exists. Power depends on the significance level, the effect size, the sample size, and the variability within each data set. Typically, studies are designed to have a power of 80%.	Evaluating statistically significant differences in the concentration of microorganisms at a single location over time or at multiple locations that differ in treatment, storage, or transport.

3.5.1 Power analysis for planning of microbiological sampling programs

From a planning perspective, power analysis can be used to determine the minimum amount of data needed to test a specific null hypothesis. Some knowledge of the intrinsic variability in the population tested is needed, such as the coefficient of variation (CV) or the standard deviation. A graphical comparison of the number of samples required for comparing two groups of data based on the percent change and coefficient of variation is shown in Figure 3-3. Larger numbers of samples are needed to detect smaller changes between groups of samples; for example, in Figure 3-3, when the CV of the data set is 0.1, only 10 samples are needed to detect an ~10% difference, but 100 samples are needed to detect an ~5% difference. These graphs also show that, the lower the variability associated with a given data set, the fewer samples are needed.

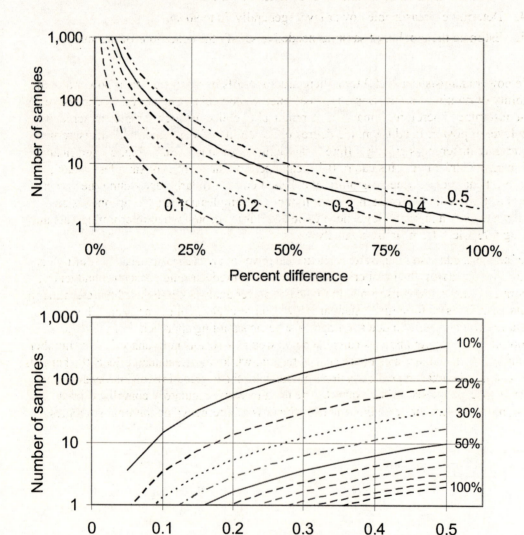

Figure 3-3. Number of samples needed to detect a statistically significant difference between samples when the coefficient of variation ranges from 0.1 to 0.5 and the percent difference ranges from 10 to 100% (calculated based on alpha level of 0.05 and beta value of 0.2).

The starting point for power analysis is based on the types of microbiological measurements to be conducted. Historical data on the variability of the microbial measurements for the type of sample to be tested can be used to estimate the standard deviation or coefficient of variation. The number of samples needed for a given application can be estimated by following the five steps listed below.

1. Decide the significance level, alpha, to be used for the analysis (acceptable false-positive error rate).

2. Determine the expected level of difference among the test groups in terms of differences in microbial concentrations, log concentrations, or percent difference.

3. Use historical data to determine the anticipated variability for a given measurement based on standard deviations or coefficient of variation.

4. Determine a reasonable power level (generally 70 to 90%).

5. Estimate the number of samples needed to satisfy the conditions of the study.

While power analysis is a useful tool, there are trade-offs between statistical power and the feasibility of running a study. Some types of microbiological testing are fairly expensive and time-consuming. Therefore, it may not be practical to collect enough samples to achieve a given level of power. In addition, the degree of variability of some microbial measurements may obscure differences among different sampling conditions. Using a higher significance level (e.g., set alpha to 0.1 instead of 0.05) can decrease the number of samples needed; however, it can increase the possibility of a Type I error. Similarly, decreasing the power level of a study will also result in a lower number of samples needed. Compromises on significance level and power level need to be taken into account when interpreting data and drawing inferences from statistical analyses.

Power analysis can also be used to evaluate the power available from a given experimental design. In some cases, the number of sample events is predetermined based on budgetary, temporal, or situational constraints. In that case, power analysis can be used to determine the detectable effect size for a given standard deviation or coefficient of variation. This calculation can be useful in determining if a specific sampling approach will yield the appropriate data. Power analysis can also be viewed as a cost–benefit analysis. The number of samples to be tested relates to the cost of testing, while the effect size reflects the benefits of the testing. Another important factor is to consider the degree of precision necessary to answer a specific question. For example, the need to verify regulatory compliance may require more power than testing of the effectiveness of alternative operational strategies.

3.5.2 Factors that impact the accuracy of power analysis

While power analysis can provide an estimate of the number of samples needed to test a specific hypothesis, there are other factors that should be considered in designing sampling programs for evaluating reclaimed water. Typically, treatment facilities are designed to meet specific discharge requirements, but the characteristics of reclaimed water can vary with the time of day, season, and weather (precipitation and wind) events. For example, in urban settings, flowrates to reclaimed water facilities vary with the patterns of water users and the effectiveness of stormwater management.

Variations in flowrates translate into variations in hydraulic retention time in individual treatment units. Typically, peak flows occur in the morning and evening hours, while low flows occur during the night. Thus, samples collected in the early morning may reflect lower flowrates, longer hydraulic residence times, and potentially more efficient treatment than samples collected under peak flow conditions. In applying the results from power analysis to determine the number of samples to be collected for a given project, it is important that the sampling program reflect consistent conditions at a treatment facility, depending on the specific goal of the sampling. Thus, when multiple samples are collected from a given sampling location, efforts should be made to minimize operational variations at a facility such as flowrates and loading rates.

CASE STUDY

BOX 7: HOW MANY SAMPLES?

To meet upcoming changes in regulatory requirements, a reclaimed water plant is seeking to determine if upgrading its filtration system will improve removal of microorganisms and be cost-effective. Preliminary screening has been used to identify three candidate approaches: use the existing system (monomedia), increase the depth of the existing filter and add a second type of medium (dual-medium filter), and install a membrane process (microfiltration). Plant managers have requested development of a pilot test to compare how well the three filtration approaches remove microorganisms. The pilot testing program will be conducted over a 6-month period, and three filters will be operated in parallel using a common influent feed. Because of concerns about the cost of the study, it is also important to have a strategy for obtaining meaningful data by analyzing the fewest number of individual samples.

In setting up the pilot-scale testing program, an estimate is needed of the **number of samples** required to determine whether there is a significant difference in microbial removal among the three filter types—otherwise, the sample size might not justify the time and expense of conducting the study.

The existing filtration system (monomedia) will be used to generate the data used for power analysis. First, data are needed on the degree to which microbiological concentrations vary before and after filtration, specifically the mean and standard deviation. To keep costs low, total coliforms will be used to "model" bacterial variability because this test is relatively inexpensive and easy to perform. *Cryptosporidium* oocysts will also be enumerated to provide additional insight into filter performance.

Seasonal variations in flowrates and water characteristics can also occur, depending on the location. It is important that sampling events intended to evaluate seasonal fluctuations are also scheduled to accommodate variations in wastewater characteristics and flowrates. Microbiological factors that are affected by seasonality include growth rates and die-off rates. In general, higher growth and die-off rates (depending upon the specific microorganism) are associated with warm weather. Increased rates of pathogen survival are associated with colder weather. Thus, the impacts of temperature on microorganism persistence need to be integrated into the overall design of the sampling program.

BOX 7a: PRELIMINARY DATA ON TOTAL COLIFORM CONCENTRATIONS

To develop preliminary data, the effluent from the existing filter is sampled 12 times over 3 weeks and the concentration of total coliforms is determined. We are interested in collecting enough samples to demonstrate a difference in effluent concentrations of at least 20 CFU/100 mL. The geometric mean and standard deviation of the data are shown in the table below.

Statistic ($n = 12$)	Log_{10}-transformed value	Geometric mean total coliform concentration, CFU/100 mL
Mean	2.04	110
Standard deviation	1.57	37

Results of the power calculation (standard deviation of 1.57) are shown in Figure 7 and itemized in Appendix 1. A summary of the number of samples and the differences that can be detected at 80% power is shown in the table below. If the study is designed to collect >25 samples, we should be able to demonstrate differences of >20 CFU/100 mL (assuming the standard deviation is consistent).

No. of samples in each set of observations	Difference that can be detected at 80% power, CFU/100 mL
12	76
20	27
25	**19**
30	14

3.5.3 Conducting a power analysis

A good starting point for determining the number of samples needed is to design a study to yield a power of 80%. This assumption means that, on average, a significant difference between the mean concentrations in two sets of observations would be found in 80 out of 100 tests, if there is a real difference.

It is valuable to base the power analysis on actual data if possible. If you already have an estimate of the standard deviation from some preliminary study (or from the literature), it can be used with the value of beta and the minimum effect size you wish to detect to estimate the power of the analysis under different sample sizes. If the number of samples is fixed (for example, by budgetary constraints), you can calculate the minimum effect size detectable, given beta and the standard deviation.

Many statistical programs can calculate these values including the free, open-source package R (http://www.r-project.org), shareware available on the Internet, or commercially available programs such as GraphPad StatMate™.

The other piece of information needed to calculate power is the alpha value. Typically, alpha values of 0.05 are used, although higher values (e.g., 0.10) could be used, depending on the context.

A comparison of the relationship between the number of samples and the minimum difference that can be detected between two sets of data is shown in Figure 3-4a for data with a standard deviation of 1.57 and an alpha value of 0.05 and in Figure 3-4b for data with a standard deviation of 0.67 (alpha = 0.05).

As shown in Figure 3-4, the lower the number of observations, the more difficult it is to detect SMALL differences in means. It is important to understand the meaning of power. If the power is 50%, that means that in half of the tests, a statistically significant difference between two groups with a given standard deviation would be found. In the other half of the tests, the difference would not be detected (a Type II

BOX 7b: PRELIMINARY DATA ON *CRYPTOSPORIDIUM* CONCENTRATIONS

To make sure that the sampling program will also be able to yield useful data on *Cryptosporidium* **concentrations,** an additional set of preliminary data was collected. Results are shown in the table below. Note that this standard deviation is considerably smaller than that calculated for the total coliform data set.

Statistic ($n = 12$)	Log_{10}-transformed value	Geometric mean *Cryptosporidium* concentration, oocysts/100 L
Mean	1.45	28
Standard deviation	0.67	4.7

No. of samples per set of observations	Difference that can be detected in oocysts/100 L	
	80% power	99% power
12	6.3	17
20	4	8.5
25	**3.5**	**6.8**
30	3	5.8

By collecting >25 samples, we should be able to demonstrate differences of >4 oocysts/100 L at 80% power and >7 oocysts/100 L at 99% power (assuming the standard deviation is consistent).

---

error). Thus, designing a study with a power of 50% is not likely to yield useful information. A higher power level will result in a lower potential error rate. To achieve more statistical power, either more samples are needed or the level of difference between the data sets that can be discriminated is much greater. We can use Figure 3-4 to compare this concept on a practical level. Figure 3-4 was generated based on log-transformed observations of microbial concentrations with a standard deviation of 1.57 ($\log_{10}$ transformed). If 10 samples are collected and the standard deviation is 1.57, then the minimum difference that can be detected on a $\log_{10}$ scale is 1.46 at 50% power, 2.09 at 80% power, and 3.19 at 90% power. Samples were tested for a) log-transformed total coliform data with a geometric mean of 110 CFU/100 mL and a standard deviation of 1.57 ($\log_{10}$ scale) at power levels ranging from 50 to 99% and b) log-transformed *Cryptosporidium* data with a geometric mean of 28 oocysts/100 L and a standard deviation of 0.67 ($\log_{10}$ scale) at power levels ranging from 50 to 99%. The lines in each graph represent power levels ranging from 50 to 99%. These differences are translated into actual concentrations in Table 3-7. This information can be used to help to frame the study. If we are trying to ask whether the survival of microorganisms (log-normally distributed with a standard deviation of 1.57) through two different types of treatment system differs, and we could collect only 10 samples, then there would need to be a difference of greater than 123 CFU/mL (80% power) between the mean values of the 10 observations from each system. However, if we had the opportunity to collect 20 data sets from each system, we could detect differences of about 27 CFU/100 mL (80% power) and 50 samples would allow for detection of differences of 8 CFU/100 mL. In each of these cases, there would be a 20% rate of Type II error. Increasing the power will decrease the error rate but require more samples. The public health significance of these differences should also be considered.

Table 3-7. Comparison of the differences in total coliform concentrations that can be detected for different numbers of samples at power levels ranging from 50 to 90% ($\log_{10}$-transformed standard deviation of 1.57)

| No. of samples | Difference in total coliform concentrations that can be detected, CFU/100 mL | | |
| | % Power | | |
	50%	80%	90%
10	29	123	1549
20		20	
50		8	

There are some general caveats that need to be considered in estimating the number of samples needed. If we use one set of preliminary data to develop an initial estimate, it is important that all microbial data do not necessarily display similar standard deviations. Therefore, the number of samples needed to demonstrate a difference between concentrations of total coliforms is likely to be different from the number of samples needed to detect differences in *Cryptosporidium* or other pathogens.

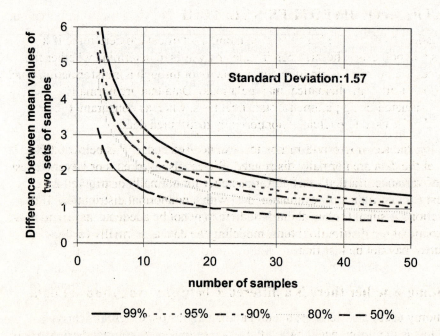

a. Total coliform data from reclaimed water facilities with a standard deviation of 1.57($\log_{10}$ scale)

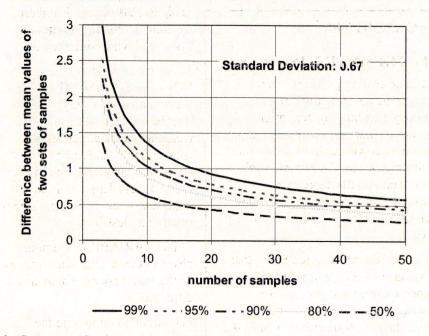

b. *Cryptosporidium* oocyst data with a standard deviation of 0.67 ($\log_{10}$ scale)

Figure 3-4. Relationship between the difference that can be detected between two groups of samples and the number of samples tested.

3.6 APPLICATIONS OF HYPOTHESIS TESTING

Most types of statistical analyses are based on evaluating a set of data to determine if it supports or refutes a hypothesis. The first step in data analysis is to determine whether the data can be modeled as a normal distribution either directly or through a transformation (such as log transformation or other mathematical manipulations). Data that are normally distributed can be evaluated using parametric statistical tests, whereas nonparametric statistical tests need to be used to evaluated nonnormally distributed data.

To test for normality, the Kolmogorov-Smirnov test can be used and, if P is below 0.05, it can be assumed that the data are normally distributed. We can use Bartlett's or Levene's test for homogeneity of variance; Bartlett's test performs better for normally distributed data, while Levene's test is more robust for evaluating data with a nonnormal distribution. If the number of observations is small (below about 10), there may not be adequate statistical power to determine the shape of the distribution; thus, modeling the data as normally (or log-normally) distributed may not be justified.

3.6.1 Determining whether there is a difference between two groups of data

A *t* **test** is a commonly used statistical test to compare the means of two groups. In its simplest form, a *t* test is the difference between two means, divided by the standard error of the difference between the means. A large value of *t* tells us that the two means are indeed likely to be different. But there are some important conditions associated with using the *t* test:

1. The data must be NORMALLY DISTRIBUTED. If the data are not normally distributed, then just using the means and standard errors for comparison would not be an adequate characterization of the two samples! This is particularly important when the sample size is small (e.g., less than 100).

2. The VARIABILITY among observations in each data set must be the same (not systematically different).

In addition to mitigating the effects of very large or very small observations on the mean, data transformation sometimes allows the data to meet the assumptions necessary for the use of parametric statistical tests, which are more powerful than nonparametric tests.

CASE STUDY

BOX 8: HYPOTHESIS TESTING: *t* Test

We are interested in determining if filtration is effective for reducing the concentration of *Cryptosporidium* oocysts. Thus, concentrations of *Cryptosporidium* oocysts must be compared before (upstream of) and after (downstream of) filtration. Based on initial power analysis, 24 paired samples were collected upstream and downstream of filtration over the course of a year. Supplementary data on filter operation, flowrates, and water quality were also collected.

A one-tailed PAIRED *t* test will be used rather than an unpaired test because the pre- and postfiltration samples are matched (paired) and because we assume that the concentration AFTER filtration will be equal to or lower than the concentration BEFORE filtration (one-tail), unless microorganisms are mobilized from the filter during filtration.

The data (shown in Case Study Box 8a) are organized in columns for prefiltration observations and postfiltration with row-wise organization of each data pair (prefiltration and postfiltration).

Another useful feature of paired *t* tests is that supplemental testing can determine the extent to which the two groups are correlated (i.e., a change in one predicts a change in the other).

Because the data must be normally distributed if one is to compare means using a parametric Student t test; the first step in data analysis is to determine if the data can be modeled as a normal distribution. A common approach is to use the Kolmogorov-Smirnov normality test to determine if $P < 0.05$. If the data are not normally distributed, then the data can be transformed and re-evaluated for normality. A typical approach is to log-transform the data by taking the $\log_{10}$ of each observation.

For data that can be modeled using a normal (or log-normal) distribution, the next step is to compare the variances of the groups to see if they are systematically different ($P < 0.05$).

The third step is to compare the sample means using a parametric Student t test.

For data that cannot be modeled using a normal (or log-normal) distribution, nonparametric data analysis tools must be employed to compare characteristics of different sets of data.

CASE STUDY

BOX 8a: DATA FOR HYPOTHESIS TESTING

Sample calculations for the removal of *Cryptosporidium* through filtration are based on the numerical and $\log_{10}$-transformed data shown below.

Numerical values, oocysts/100 L		$\log_{10}$-transformed, log (oocysts/100 L)	
Prefiltration	Postfiltration	Pre-filtration	Post-filtration
222	61	2.3	1.8
259	196	2.4	2.3
228	112	2.4	2.0
10	3.5	1.0	0.5
31.7	10.6	1.5	1.0
17.6	3.9	1.2	0.6
31	5.9	1.5	0.8
61.2	2.2	1.8	0.3
179	8.6	2.3	0.9
103	14	2.03	1.1
18.3	3.9	1.3	0.6
10	100	1.0	2.0
13.7	157	1.1	2.2
615	2.8	2.8	0.4
12.8	4.1	1.2	0.6
679	4.1	2.8	0.6
27	59	1.4	1.8
345	275	2.5	2.4
21.2	5.3	1.3	0.7
10.6	5.2	1.0	0.7
10.6	10.6	1.0	1.0
84	12	1.9	1.1
18	1	1.3	0
42	42.3	1.6	1.6

3.6.2 ANOVA

If more than two sets of data are to be compared, then an ANOVA can be conducted. ANOVA, like the *t* test, is designed to tell us whether the differences between means of groups are statistically significant. Also like the *t* test, the assumptions about the data include that they are normally distributed and that variances are homogeneous (not systematically different).

A one-factor ANOVA can be used to compare three sets of data based on a common factor, such as degree of treatment, time of day, season, etc. The test statistic for an ANOVA is the *F* statistic, where

$$F = \frac{Variance\,Among\,Treatments}{Variance\,Within\,Treatments}$$

If the null hypothesis that treatments have no effect is true, the two variances should be about the same; if treatments do have an effect, $F \gg 1$. Because there are two variances involved, an *F* test has degrees of freedom (df) for both the numerator and denominator. For example, if there are 5 numerator df and 100 denominator df, *F* is usually written as $F_{5,100}$. If the *F* test shows a significant difference among treatments, then post-hoc tests can be conducted to gain more insight into the differences among the data sets by comparing two groups of data (similar to the *t* test).

CASE STUDY

BOX 9: ANOVA

The effectiveness of three different filters for removal of microorganisms will be compared using ANOVA. The three filters include monomedia, a dual-medium filter, and a membrane process (microfiltration). A 6-month pilot testing program was designed to test the three filters in parallel using a common influent feed. This design minimizes the variability in influent delivered to the filters: one influent stream is split three ways, so that each filter receives comparable influent and hydraulic loading rates.

A one-factor ANOVA will be conducted, and the filter type is the factor which has three levels: monomedia, microfiltration, and dual media. For this example, we will consider the effect of filtration on the concentration of total coliforms. The $\log_{10}$-transformed data are provided in Appendix 2.

Based on the Kolmogorov-Smirnov test, $P < 0.05$ for two of the three groups (microfiltration and dual media), indicating that the distribution of the data is not Gaussian (normal). We can use Bartlett's or Levene's test for homogeneity of variance; Bartlett's test performs better with normally distributed data, while Levene's test is more robust for data from a nonnormal

3.6.3 Multivariate data analysis

Certain kinds of data are inherently *multivariate*—that is, we measure several things simultaneously on the same unit. For example, in comparing the efficacy of several types of filters, we might measure, for each replicate of each type of filter, concentrations of total coliforms and *Cryptosporidium* oocysts after filtration (along with other water quality and operational data). In general it is important to think of each set of measurements as being a single multivariate observation. One reason to use multivariate analysis is that the quantities we are measuring may be correlated with one another (positively or negatively). Treating the entire set of measurements (technically, a vector of measurements) as an observation allows us to conduct multivariate statistical analyses that can account for these correlations. It is possible, for example, that a multivariate test might show that our filters differ from one another even though univariate tests (like standard ANOVA) conducted separately on coliforms and *Cryptosporidium* do not. The reverse is also true.

One of the most useful multivariate methods is the multivariate analysis of variance (MANOVA). Many statistical packages can perform the necessary calculations. MANOVA calculates several other quantities that are analogous to the F tests used in ANOVA. There are standard formulae (built in to the software) that translate these quantities into F statistics for significance tests. There are four different MANOVA test statistics that are analogous to F statistics, and in many cases all four will give roughly the same answer. We recommend using Pillai's trace, since it is the most robust.

If results from MANOVA suggest that there are significant differences among the test variables, it is possible to conduct post-hoc multivariate comparisons. These comparisons use the pooled error data (as a post-hoc Tukey test uses the pooled error variance) and are consequently very powerful. See Appendix 2 for an example of the output from a MANOVA test.

CASE STUDY

BOX 10: MANOVA

MANOVAs can be used to answer the question: are there differences among the filters in the multivariate (total coliforms and *Cryptosporidium*) concentrations of microorganisms? MANOVA can be performed in almost all general statistics packages; results from the SAS analysis are summarized here

For the filter data, Pillai's trace (with 4,174 df) has a P of <0.0001, suggesting that differences between filters are quite strong.

Post-hoc multivariate comparisons.

Post-hoc comparisons use the pooled error data (as a post-hoc Tukey test uses the pooled error variance) and are consequently very powerful.

Comparison	P
A vs. C	<0.0001
B vs. C	<0.0001
A vs. B	0.96

Based on these P values, we can conclude that filter C (the membrane systems) performs differently from both of the other filters and that the other two filters (mono- and dual media) are not distinguishable from one another.

It should be noted that differences in P do not translate into the relative importance of each comparison. Other types of tests are needed for this analysis (see Box 11).

As a follow-up to MANOVA, one can perform a canonical analysis to obtain estimates for the relative importance of effects. While the term "canonical analysis" may sound forbidding, the idea is straightforward and closely related to discriminant function analysis. The goal is to identify which terms in the model account for most differences among groups. The SAS software can readily perform this analysis.

CASE STUDY

BOX 11: POST-HOC ANOVA TESTING

Based on the MANOVA results, follow-up statistical questions can be posed:

- Are the observed statistically significant differences among filters due to their different effects on total coliforms, *Cryptosporidium*, or both?

- What is the relative importance of each species in accounting for the observed difference among the filters?

First, we can conduct univariate ANOVAs for each species.

Microbial analyte	F	P
Total coliforms	2.60	0.08
Cryptosporidium	19.73	<0.0001

These results suggest that the differences among filters are significant for *Cryptosporidium* but are not quite significant for total coliforms.

To quantify the relative importance of each species for the differences among filters, we can calculate the canonical coefficients from MANOVA (from SAS).

Microbial analyte	Standardized canonical coefficient
Total coliforms	0.017
Cryptosporidium	1.18
Ratio	69

This analysis tells us that *Cryptosporidium* is about 69 times more important than total coliforms in accounting for the differences among filters.

3.6.4 Correlation analysis

Correlation analysis is used to determine the strength of the linear relationship between two variables. For example, it might be used to determine whether the concentration of bacteriophages in effluent samples is correlated with the concentration of enteric viruses or whether total coliform concentrations are correlated with fecal coliform concentrations in filtered effluent. When two parameters are correlated, it means that change in one quantity is usually accompanied by a change in the other. Parameters may be positively correlated (as A increases, so does B) or negatively correlated (e.g., as A increases, B decreases). Correlation does not imply causation, but it can be used to help interpret data and estimate parameters.

The Pearson r correlation can be used to analyze normally distributed data. The statistics used to assess the strength of correlations include P, which indicates the probability that a correlation coefficient as large as the one obtained would occur by chance, and Pearson's r, the correlation coefficient. The smaller P is and the larger r is (closer to 1.0 or −1.0), the better the correlation.

An example of a correlation of paired data from total coliform and fecal coliform measurements ($\log_{10}$ transformed) of filtered reclaimed water is shown in Figure 3-5. For the data in Figure 3.5, the $\log_{10}$-transformed $P < 0.001$ and the Pearson r is 0.87. The positive r value and slope mean that, as the independent variable increases, so does the dependent variable. Nonnormally distributed data can be analyzed in a similar manner using the Spearman ranked correlation. Further analysis can be conducted to generate an equation that describes the correlated data.

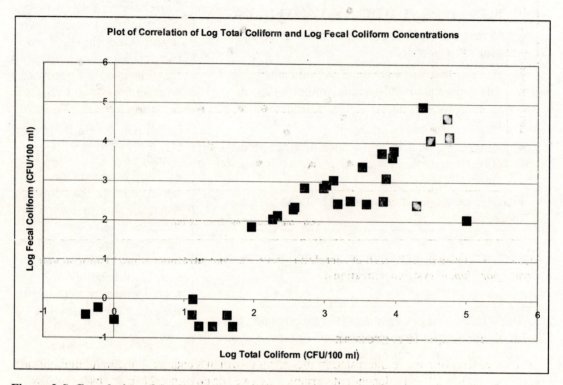

Figure 3-5. Correlation of total coliform and fecal coliform concentrations.

Another example of an attempt to correlate data is shown in Figure 3-6, where $P = 0.41$ and Pearson's $r = 0.15$. In this case, $\log_{10}$-transformed total coliform concentrations are compared to *Cryptosporidium* oocyst concentrations ($\log_{10}$ transformed).

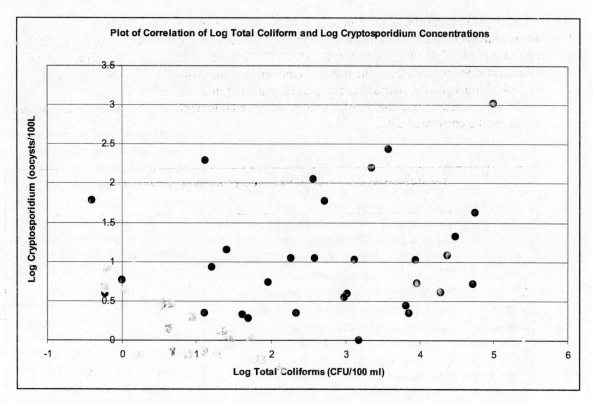

Figure 3-6. Evidence of the lack of correlation between concentrations of total coliform and *Cryptosporidium* oocyst concentrations.

3.6.5 Binary logistic regression

In some cases, we may be interested in the extent to which a change in the magnitude of one variable predicts a binary dependent variable (e.g., yes or no or presence or absence). For example, one might want to determine whether the measured concentration of an indicator bacterium type is correlated with the probability of detecting a specific pathogen. Such questions are particularly pertinent when analyzing data in which the desired goal is to produce a pathogen-free product, such as in disinfection of reclaimed water. Binary (or binomial) logistic regression can be used to analyze data for such purposes. The analysis can also accommodate binary dependent and independent variables (e.g., presence or absence of both indicator and pathogen) and multiple independent variables.

Binary logistic regression uses maximum likelihood estimation to determine the probability that a certain event will occur, for example, detection of the specific pathogen. The output of this analysis is the change in the log odds of the dependent variable. The change in log odds may be negative (i.e., odds of detection go DOWN as the magnitude of the independent variable increases) or positive (i.e., odds of detection go UP as the magnitude of the independent variable increases). A pseudo-R^2 statistic demonstrates the strength of the relationship between the independent and dependent variables. Binary logistic regression does not assume a linear relationship between variables and does not assume a normal distribution of the data; therefore, it is useful for many data sets that are problematic for analysis with other statistical methods.

The outputs of binary logistic regression in a statistical program such as SPSS include the model chi-square test. It is the most commonly used test of significance for this method and should be significant at the alpha level of 0.05 or better if the variables are significantly correlated. Although the Wald statistic is also frequently used to test the significance of each independent variable in multiple regressions, it is prone to Type II errors (finding that the variable is not significant when it really is). For the strongest relationships, both the Wald statistic and the model chi-square will be < 0.05. The R^2 value that is most reported for binary logistic regression is Nagelkerke's R^2, which is always larger than the Cox and Snell's R-Square from which it is derived. Possible values of Nagelkerke's R^2 range from 0 to 1, with 0 representing no correlation and 1 representing a perfect correlation.

The odds ratio [Exp(b) in SPSS] is another pertinent value for binary logistic regression. Basically, the odds ratio expresses the likelihood that a one-unit change in the independent variable will correspond to a specific outcome for the dependent variable. If, for example, the concentration of coliforms is the independent variable and the presence or absence of *Cryptosporidium* is the dependent variable, then there are three possibilities as outlined in Table 3-8: the odds ratio could be 1.0, <1.0, or >1.0, representing no correlation, negative correlation, or positive correlation, respectively (see Table 3-8). The 95% confidence interval around the odds ratio can be used to assess the significance of the correlation; if the 95% confidence interval includes 1.0, the correlation is generally not considered significant.

Table 3-8. Interpretation of different odds ratios

Odds ratio	Interpretation
> 1.0	The independent and dependent variables are positively correlated; i.e., a change in the independent variable indicates the dependent variable will change in a similar way. In binary logistic regression, the only option for the dependent variable is a categorical chain, i.e., +/− or 1/0.
1.0	Change in the independent variable IS NOT related to change in the dependent variable.
< 1.0	The independent and dependent variables are negatively correlated; i.e., a change in the independent variable indicates the dependent variable will change in the opposite way (if the independent variable increases, the dependent variable will decrease and vice versa).

CHAPTER 4

DATA MANAGEMENT

Because of the nature of most types of microbial analyses, there is a need to manually enter microbiological test results into spreadsheets or databases. While use of spreadsheets is fairly commonplace in the reclaimed water community, preliminary planning of criteria for entering microbiological data can facilitate statistical analyses. Key issues that need to be considered are:

- Inclusion of categorical data

- Appropriate units for numerical data

- Management of "nondetect" or censored data

- Transparency of database structure: will someone else be able to understand your database and find information without frustration?

- Consistency of database format: will another computer program be able to read in your spreadsheet (or other file) for analysis?

4.1 USE OF DATABASES TO INTEGRATE MICROBIOLOGICAL DATA WITH PROCESS DATA AND OTHER MONITORING DATA

Databases are computer-based tools designed to facilitate organization, sorting, and analysis of information. A variety of database packages are available both as commercial and free open-source products, and they vary in complexity in terms of the ease with which data can be entered, retrieved, and analyzed. In general, databases are designed for relatively easy retrieval of data as long as they are entered and stored in an appropriate and consistent format. A major benefit of using databases for managing data pertaining to reclaimed water is that comprehensive data can be readily accessed for conduct of retrospective studies or time-series analyses. Another advantage of using databases is that the integration of process and water quality information with microbiological data can be streamlined through the development of targeted datum queries.

4.2 SPREADSHEET CONSIDERATIONS

Spreadsheets are widely used for storage and communication of data. Spreadsheets allow for tabular entry of information, relatively easy graphical analysis, and statistical calculations. Spreadsheets can be designed to allow for querying of data and comparing of various data sets, but these tasks can be more cumbersome than the use of databases, depending on the amount of information that needs to be processed. Spreadsheet formats are often designed for ease of use. However, most statistical analysis and graphics programs need data in a different format. Putting such data into a readable format sometimes requires considerable effort and programming. It is almost always more cost-effective to pay a programmer at the outset of a project to design a database that will be easily accessible for statistical analysis than to set up a spreadsheet for ease of use and then have to pay a programmer to translate the data into a format that can be analyzed.

When entering microbiological data in spreadsheets or databases, it is important to format the data properly to allow for conducting statistical calculations. A consistent form of data, either numerical or alphanumeric, should be stored in each column. Quantitative information should be stored as numerical data, and the units or detection limits associated with each sample should be stored in a column adjacent to the numerical data. In addition, categorical information should be stored in a column separate from numerical data.

CHAPTER 5
SUMMARY AND CONCLUSIONS

This guidance manual was developed to provide users with a context for collecting, exploring, and interpreting microbiological data associated with reclaimed water. Basic concepts are presented to facilitate collection of meaningful information with an emphasis on design of sampling programs, data interpretation, and statistical analysis. The information provided here can help with routine monitoring programs and design of studies for detailed microbiological investigations. The examples and illustrations provided in this document are intended to help the reader tackle a range of microbial investigations associated with reclaimed water facilities. More advanced and detailed information can be found in numerous microbiological and statistical reference books. A suggested reading list is provided in Appendix 3.

To recap some of the important points considered in this document:

- Frame your questions carefully, with consideration of available supporting data and resources.

- Plan your experimental design with system characteristics and specific goals in mind; for example, are analyses to be paired or unpaired; can many samples be analyzed or are there cost considerations that will necessitate compromises?

- Obtain preliminary data; carry out descriptive statistics and make graphs to gain an understanding of the variability and other characteristics of the data.

- Determine the expected necessary sample size given the variability and the magnitude of the difference you wish to detect.

- Have fun with statistical calculations and interpretation of results! Don't be afraid to try using different statistical approaches. Once you start manipulating data and seeing relationships within and among data sets, it can become an absorbing and very helpful part of your skill set.

REFERENCES

Data Quality Assessment: Statistical Methods for Practitioners; EPA/240/B-06/003; U.S. Environmental Protection Agency, U.S. Government Printing Office: Washington, DC, 2006.

Dupont, H.L.; Chappell, C.L.; Sterling, Q.R.; Okhuysen, P.C.; Rose, J.B.; and Jakubowski, W. The infectivity of *Cryptosporidium* parvum in healthy volunteers. *N. Engl. J. Med.* **1995,** *332,* 855–859.

Edwards, A. W. F. *Likelihood;* Baltimore: Johns Hopkins University Press, 1992.

Guidance on Systematic Planning using the Data Quality Objectives Process; EPA/240/B-06/001; U.S. Environmental Protection Agency, U.S. Government Printing Office: Washington, DC, 2006.

Harwood, V. J.; Levine, A. D.; Scott, T. M.; Chivukula, V.; Lukasik, S. R.; Farrah, S. R.; Rose, J. B. Validity of the indicator organism paradigm: pathogen reduction and public health protection in reclaimed water. *Appl. Environ. Microbiol.* **2005,** *71,* 3163–3170.

Littell, R. C.; Milliken, G. A.; Stroup, W. W.; Wolfinger, R. D. *SAS System for Mixed Models;* SAS Institute: Cary, NC, 1996.

Lorimer, M. F.; Kiermeier, A. Analysing microbiological data: Tobit or not Tobit? *Int. J. Food Microbiol.* **2007,** *116,* 313–318.

Pinheiro, J. C.; Bates, D. M. *Mixed-Effects Models in S and S-PLUS;* Statistics and Computing Series; Springer-Verlag: New York, 2000.

Standard Methods for the Examination of Water and Wastewater, 21st ed.; Eaton, A. D., Clesceri, L. S., Rice, E. W., Greenberg, A. E., Franson, M. A. H., Eds.; American Public Health Association: Washington, DC, 2005.

APPENDICES

APPENDIX 1
POWER ANALYSIS

Output from power analysis: total coliform data; standard deviation is 1.57

N per group	Power				
	99%	95%	90%	80%	50%
3	6.99	5.88	5.29	4.57	3.20
4	5.62	4.73	4.25	3.67	2.57
5	4.84	4.07	3.66	3.16	2.21
6	4.31	3.62	3.26	2.82	1.97
7	3.92	3.30	2.97	2.56	1.79
8	3.63	3.05	2.74	2.37	1.66
9	3.39	2.85	2.56	2.21	1.55
10	3.19	2.68	2.41	2.09	1.46
12	2.88	2.42	2.18	1.88	1.32
14	2.65	2.23	2.00	1.73	1.21
16	2.47	2.07	1.86	1.61	1.13
18	2.32	1.95	1.75	1.51	1.06
20	2.19	1.84	1.66	1.43	1.00
25	1.95	1.64	1.47	1.27	0.89
30	1.77	1.49	1.34	1.16	0.81
35	1.63	1.37	1.24	1.07	0.75
40	1.53	1.28	1.15	1.00	0.70
50	1.36	1.14	1.03	0.89	0.62
60	1.24	1.04	0.94	0.81	0.57
70	1.15	0.96	0.87	0.75	0.52
80	1.07	0.90	0.81	0.70	0.49
90	1.01	0.85	0.76	0.66	0.46
100	0.96	0.80	0.72	0.63	0.44
150	0.78	0.66	0.59	0.51	0.36
200	0.67	0.57	0.51	0.44	0.31
300	0.55	0.46	0.42	0.36	0.25
400	0.48	0.40	0.36	0.31	0.22
500	0.43	0.36	0.32	0.28	0.19

Output from power analysis: *Cryptosporidium* data; standard deviation is 0.67

N per group	Power				
	99%	95%	90%	80%	50%
3	2.98	2.51	2.26	1.95	1.36
4	2.40	2.02	1.81	1.57	1.10
5	2.06	1.74	1.56	1.35	0.94
6	1.84	1.55	1.39	1.20	0.84
7	1.67	1.41	1.27	1.09	0.77
8	1.55	1.30	1.17	1.01	0.71
9	1.45	1.22	1.09	0.95	0.66
10	1.36	1.15	1.03	0.89	0.62
12	1.23	1.03	0.93	0.80	0.56
14	1.13	0.95	0.86	0.74	0.52
16	1.05	0.88	0.80	0.69	0.48
18	0.99	0.83	0.75	0.65	0.45
20	0.93	0.79	0.71	0.61	0.43
25	0.83	0.70	0.63	0.54	0.38
30	0.76	0.64	0.57	0.49	0.35
35	0.70	0.59	0.53	0.46	0.32
40	0.65	0.55	0.49	0.43	0.30
50	0.58	0.49	0.44	0.38	0.27
60	0.53	0.45	0.40	0.35	0.24
70	0.49	0.41	0.37	0.32	0.22
80	0.46	0.38	0.35	0.30	0.21
90	0.43	0.36	0.33	0.28	0.20
100	0.41	0.34	0.31	0.27	0.19
150	0.33	0.28	0.25	0.22	0.15
200	0.29	0.24	0.22	0.19	0.13
300	0.23	0.20	0.18	0.15	0.11
400	0.20	0.17	0.15	0.13	0.09
500	0.18	0.15	0.14	0.12	0.08
1000	0.13	0.11	0.10	0.08	0.06

APPENDIX 2

MANOVA OUTPUT

SAS output for MANOVA. Analysis of filter effect data.

The GLM Procedure

MANOVA Test Criteria and F
Approximations for the Hypothesis of No Overall Filter Effect

H = Type III SSCP Matrix for filter
E = Error SSCP Matrix

Statistic	Value	F Value	Num DF	Den DF	Pr > F
Wilks' lambda	0.68793678	8.84	4	172	<0.0001
Pillai's trace	0.31207067	8.04	4	174	<0.0001
Hotelling-Lawley trace	0.45361112	9.72	4	102	<0.0001
Roy's greatest root	0.45358724	19.73	2	87	<0.0001

Canonical analysis from the same output:

Test of H0: The canonical correlations in the current row and all that follow are zero

Parameter	Eigenvalue	
	1	2
Difference	0.4536	0
Proportion	0.4536	
Cumulative	0.9999	0.0001
Likelihood Ratio	0.9999	1
Approximate F Value	0.687937	0.999976
Num DF	8.84	1
Den DF	4	87
Pr > F	<0.0001	0.9638

Canonical Structure

	Total		Between		Within	
	Can 1	Can 2	Can 1	Can 2	Can 1	Can 2
crypto	0.9999	0.0131	1.0000	−0.0001	0.9999	−0.0157
coliform	0.4251	0.9051	0.9998	0.0186	0.3630	0.9318

Canonical Coefficients

	Standardized		Raw	
	Can 1	Can 2	Can 1	Can 2
crypto	1.18485722	−0.46155448	2.30022594	−0.89604011
coliform	0.01709794	1.08566237	0.01400270	0.88912521

The contrast for A vs. B

H = Contrast SSCP Matrix for A vs. B

	crypto	coliform
crypto	0.0152752932	0.0308926826
coliform	0.0217231394	0.0308926826

Canonical Analysis
H = Contrast SSCP Matrix for A vs. B
E = Error SSCP Matrix

	Canonical Correlation	Adjusted Canonical Correlation	Approximate Standard Error	Squared Canonical Correlation
1	0.031113	−0.151151	0.106497	0.000968

WateReuse Foundation

Test of H0: The canonical correlations in the current row and all that follow are zero

Parameter	Value
Eigenvalue	1
Difference	0.001
Proportion	1
Cumulative	1
Likelihood Ratio	0.99903
F Value	0.04
Num DF	2
Den DF	86
Pr > F	0.9592

Canonical Structure

	Total Can 1	Between Can 1	Within Can 1
crypto	0.9896	1.0000	0.9850
coliform	0.5396	1.0000	0.5046

Canonical Coefficients

	Standardized Can 1	Raw Can 1
crypto	1.09781871	2.13125348
coliform	0.18713585	0.15325870

MANOVA Test Criteria and Exact F Statistics for the Hypothesis of No Overall A vs. B Effect
H = Contrast SSCP Matrix for A vs. B
E = Error SSCP Matrix

Statistic	Value	F Value	Num DF	Den DF	Pr > F
Wilks' lambda	0.99903196	0.04	2	86	0.9592
Pillai's trace	0.00096804	0.04	2	86	0.9592
Hotelling-Lawley trace	0.00096897	0.04	2	86	0.9592
Roy's greatest root	0.00096897	0.04	2	86	0.9592

The contrast for A vs. C:

	crypto	coliform
crypto	5.8082977633	5.9102936183
coliform	5.9102936183	6.0140805582

Canonical Analysis

H = Contrast SSCP Matrix for A vs. C
E = Error SSCP Matrix

	Canonical Correlation	Adjusted Canonical Correlation	Approximate Standard Error	Squared Canonical Correlation
1	0.513250	0.507244	0.078519	0.263425

Test of H0: The canonical correlations in the current row and all that follow are zero

Parameter	Value
Eigenvalue	1
Difference	0.3576
Proportion	1
Cumulative	1
Likelihood Ratio	0.73657475
F Value	15.38
Num DF	2
Den DF	86
Pr > F	<0.0001

Canonical Structure

	Total Can 1	Between Can 1	Within Can 1
crypto	0.9999	1.0000	0.9998
coliform	0.4279	1.0000	0.3665

Canonical Coefficients

	Standardized Can 1	Raw Can 1
crypto	1.18311421	2.29684214
coliform	0.02117802	0.01734417

MANOVA Test Criteria and Exact F Statistics for the Hypothesis of No Overall A vs. C Effect
H = Contrast SSCP Matrix for A vs. C
E = Error SSCP Matrix

Statistic	Value	F Value	Num DF	Den DF	Pr > F
Wilks' lambda	0.73657475	15.38	2	86	<0.0001
Pillai's trace	0.26342525	15.38	2	86	<0.0001
Hotelling-Lawley trace	0.35763546	15.38	2	86	<0.0001
Roy's greatest root	0.35763546	15.38	2	86	<0.0001

The contrast for B vs. C:

H = Contrast SSCP Matrix for B vs. C

	crypto	coliform
crypto	5.2278434153	5.2053249333
coliform	5.2053249333	5.1829034477

Canonical Analysis
H = Contrast SSCP Matrix for B vs. C
E = Error SSCP Matrix

	Canonical Correlation	Adjusted Canonical Correlation	Approximate Standard Error	Squared Canonical Correlation
1	0.493420	0.486829	0.080647	0.243463

Test of H0: The canonical correlations in the current row and all that follow are zero

Eigen-value	Difference	Proportion	Cumulative	Likelihood Approximate Ratio	F Value	Num DF	Den DF	Pr > F
1	0.3218	1.0000	1.0000	0.75653710	13.84	2	86	<0.0001

Parameter	Value
Eigenvalue	1
Difference	0.3218
Proportion	1
Cumulative	1
Likelihood Ratio	0.75653710
F Value	13.84
Num DF	2
Den DF	86
Pr > F	<0.0001

Canonical Structure

	Total Can 1	Between Can 1	Within Can 1
crypto	1.0000	1.0000	0.9999
coliform	0.4216	1.0000	0.3586

Canonical Coefficients

	Standardized Can 1	Raw Can 1
crypto	1.18698744	2.30436144
coliform	0.01205705	0.00987437

MANOVA Test Criteria and Exact F Statistics for the Hypothesis of No Overall B vs. C Effect
H = Contrast SSCP Matrix for B vs. C
E = Error SSCP Matrix

Statistic	Value	F Value	Num DF	Den DF	Pr > F
Wilks' lambda	0.75653710	13.84	2	86	<0.0001
Pillai's trace	0.24346290	13.84	2	86	<0.0001
Hotelling-Lawley trace	0.32181225	13.84	2	86	<0.0001
Roy's greatest root	0.32181225	13.84	2	86	<0.0001

APPENDIX 3

SUPPLEMENTAL SOURCES OF INFORMATION

Bitton, G. *Wastewater Microbiology,* 2nd ed.; Wiley-Liss, New York, 1999.

Haas, C. N. Microbial sampling: is it better to sample many times or use large samples? *Water Sci. Technol.* **1993,** *27,* 19–25.

Haas, C. N.; Scheff, P. A. Estimation of averages in truncated samples. *Environ. Sci. Technol.* **1990,** *24,* 912–919.

Handbook of Water and Wastewater Microbiology; Mara, D., Horan, N., Eds.; Academic Press: New York, 2003; pp 193–208.

Helsel, D. R. Less than obvious: statistical treatment of data below the detection limit. *Environ. Sci. Technol.* **1990,** *24,* 1767–1774.

Hurst, C.; R. Crawford, R.; Garland, J.; Lipson, D.; Mills, A.; Stetzenbach, L. *Manual of Environmental Microbiology,* 3rd ed.; Blackwell Publishing: Hoboken, NJ, 2007.

ICR Microbial Laboratory Manual; EPA/600/R-95/178; U.S. Environmental Protection Agency, U.S. Government Printing Office: Washington, DC, 1996.

Leclerc, H.; Mossel, D. A.; Edberg, S. C.; Struijk, C. B. Advances in the bacteriology of the coliform group: their suitability as markers of microbial water safety. *Annu. Rev. Microbiol.* **2001,** *55,* 201–234.

Levine, A. D.; Asano, T. Recovering sustainable water from wastewater. *Environ. Sci. Technol.* **2004,** *38,* 201A–208A.

Lindsey, J. K. *Introduction to Applied Statistics: A Modeling Approach;* Oxford University Press: New York, 2004.

Manual—Guidelines for Water Reuse; USEPA/625/R-92/004; U.S. Environmental Protection Agency, U.S. Government Printing Office: Washington, DC, 1992.

Method 1600: Enterococci in Water by Membrane Filter using Membrane-Enterococcus Indoxyl-β-Glucoside Agar (mEI); EPA/821-R02-022; U.S. Environmental Protection Agency, Office of Water, U.S. Government Printing Office: Washington, DC, 2002.

Method 1603: Escherichia coli (E. coli) *in Water by Membrane Filtration using Modified Membrane-Thermotolerant* E. coli *Agar (Modified mTEC);* U.S. Environmental Protection Agency, U.S. Government Printing Office: Washington, DC, 2002.

Method 1623: Cryptosporidium *and* Giardia *in Water by Filtration/IMS/FA;* EPA/821/R-99/006; U.S. Environmental Protection Agency, Office of Water, U.S. Government Printing Office: Washington, DC, 1999.

Motulsky, H. *Intuitive Biostatistics;* Oxford University Press: New York, 1995.

Murray, P.; Rosenthal, K. S.; Kobayashi, G. S.; Pfaller, M. A. *Medical Microbiology;* Mosby-Year Book, Inc.: St. Louis, MO, 2001.

National Research Council. *Issues in Potable Reuse;* National Academy Press: Washington, DC, 1998.

Quinn, G. P.; Keough, M. J. *Experimental Design and Data Analysis for Biologists;* Cambridge University Press: Cambridge, U.K., 2002.

Rose, J. B.; Dickson, L. J.; Farrah, S. R.; Carnahan, R. P. Removal of pathogenic and indicator microorganisms by a full-scale water reclamation facility. *Water Res.* 1996, *30,* 2785–2797.

Rose, J. B.; Farrah, S. R.; Friedman, D. E.; Riley, K.; Hamann, C. L.; Robbins, M. Public health evaluation of advanced reclaimed water for potable applications. *Water Sci. Technol.* 1999, *40,* 247–252.

Rose, J. B.; Huffman, D. E.; Riley, K.; Farrah, S. R.; Lukasik, J. O.; Harman, C. L. Reduction of enteric microorganisms at the Upper Occoquan Sewage Authority water reclamation plant. *Water Environ. Res.* 2001, *73,* 711–720.

Sobsey, M. D. Inactivation of health-related microorganisms in water by disinfection processes. *Water Sci. Technol.* 1989, *21,* 179–195.

Standard Methods for the Examination of Water and Wastewater, 21st ed.; Eaton, A. D., Clesceri, L. S., Rice, E. W., Greenberg, A. E., Franson, M. A. H., Eds.; American Public Health Association: Washington, DC, 2005.

Steidl, R. J.; Thomas, L. Power Analysis and Experimental Design. In Scheiner, S. M., Gurevitch, J., Eds.; *Design and Analysis of Ecological Experiments,* 2nd ed.; Oxford University Press: New York, 2002, pp 14–36.

York, D. W.; Walker-Coleman, L. Pathogen standards for reclaimed water. *Water Environ. Technol.* 2000, *12,* 59–61.